Two-Dimensional Materials for Electromagnetic Shielding

Two-Dimensional Materials for Electromagnetic Shielding

Chong Min Koo, Pradeep Sambyal, Aamir Iqbal, Faisal Shahzad, and Junpyo Hong

WILEY-VCH

Authors

Prof. Chong Min Koo
Korea Institute of Science and Technology
Materials Architecturing Research
Center
Hwarangro 14-gil 5
Seongbuk-gu
02792 Seoul
South Korea

Division of Nano & Information
Technology, KIST School, University of
Science and Technology
02792 Seoul
South Korea

KU-KIST Graduate School of
Converging Science and Technology
Korea University
02841 Seoul
South Korea

Dr. Pradeep Sambyal
Korea Institute of Science and Technology
Materials Architecturing Research
Center
Hwarangro 14-gil 5
Seongbuk-gu
02792 Seoul
South Korea

Dr. Aamir Iqbal
Korea Institute of Science and Technology
Materials Architecturing Research
Center
Hwarangro 14-gil 5
Seongbuk-gu
02792 Seoul
South Korea

Division of Nano & Information
Technology, KIST School, University of
Science and Technology
02792 Seoul
South Korea

Prof. Faisal Shahzad
Pakistan Institute of Engineering and
Applied Sciences (PIEAS)
National Center for Nanotechnology
Lehtrar Road, Nilore
45650 Islamabad
Pakistan

Dr. Junpyo Hong
Korea Institute of Science and Technology
Hwarangro 14-gil 5
Seongbuk-gu
02792 Seoul
South Korea

Cover Image: Courtesy of the authors

Library of Congress Card No.:
applied for

British Library Cataloguing-in-Publication Data
A catalogue record for this book is available from the British Library.

Bibliographic information published by the Deutsche Nationalbibliothek
The Deutsche Nationalbibliothek lists this publication in the Deutsche Nationalbibliografie; detailed bibliographic data are available on the Internet at <http://dnb.d-nb.de>.

Print ISBN: 978-3-527-34842-8
ePDF ISBN: 978-3-527-82980-4
ePub ISBN: 978-3-527-82981-1
oBook ISBN: 978-3-527-82982-8

Typesetting Straive, Chennai, India
Printing and binding CPI Group (UK) Ltd, Croydon, CR0 4YY

Printed on acid-free paper

C109617_070121

Contents

Preface

The advancement of electronic technology is inevitably associated with the spread of electromagnetic terrorism, as all electrical and electronic devices, including personal, public, and military equipment, operate by either transmitting or receiving electromagnetic waves (EMWs). These EMWs can interfere with each other through electromagnetic induction in close proximity and in integrated circuits, which is called electromagnetic interference (EMI). This phenomenon affects the performance of electronic circuits and shortens their lifespan; it can be used to hijack a country's defense system (thus risking security and sovereignty), and it is injurious to human health. Owing to these hazards, the undesirable EMWs require appropriate shielding in electrical and electronic appliances.

The principles and methods of electromagnetic shielding have been investigated for over 70 years. For shielding purposes, highly conductive metals that strongly reflect incident EMWs have been used for decades. However, as these reflected EMWs can be a source of secondary pollution, proper shielding may not be fully achieved. In the era of emerging fifth-generation (5G) technology, electrical and electronic circuits are being made smaller and smarter with compact infrastructure. These advanced portable electronics require novel shielding materials that are ultralight with a minimal thickness, low cost, and, most importantly, easy to process. Despite their high electrical conductivities, metals suffer from high densities and poor corrosion resistance and also require complex processing techniques. These challenges necessitate the development of new EMI shielding materials that minimize the drawbacks of highly conductive metals. Interestingly, two-dimensional (2D) materials, as exemplified by graphene, have emerged as promising alternatives with outstanding EMI shielding properties.

Since the discovery of graphene in 2004, 2D materials have attracted significant research attention owing to their unique chemical, mechanical, electrical, and optoelectronic properties. In addition to graphene, transition metal carbides/nitrides/carbonitrides (MXenes), transition metal dichalcogenides (TMDCs), black phosphorus (BP), hexagonal boron nitride (h-BN), and metal–organic frameworks (MOFs) possess varying electrical conductivities, dielectric permittivities, and magnetic permeabilities, coupled with other attractive physiochemical properties. Owing to their excellent electrical conductivity, processability, corrosion resistance, and low density, 2D materials are believed to be an alternative to highly conductive metals for EMI shielding. In particular, monolayer graphene and $Ti_3C_2T_x$ MXene show outstanding EMI shielding potential because of their very high electrical conductivities in nanometer-thick films. Additionally, as 2D materials can be used to make highly processable liquid dispersions, inks, low-percolation polymer composites with mechanical flexibility and light weight, these materials can satisfy both on-ground and space application requirements.

This handbook covers all fundamental aspects of EMI shielding. Chapter 1 provides a basic introduction and describes EMW sources, the potential hazards of EMWs, EMI shielding standards for ensuring electromagnetic compatibility, and the advanced 2D materials developed in this field. Chapter 2 provides a comprehensive overview of the fundamental EMI shielding mechanisms and highlights the properties required to enhance shielding effectiveness (SE). Chapter 3 briefly explains the various measurement and calculation methods used in this field. Highlighting the importance of 2D materials, the EMI shielding properties of graphene, MXenes, and other 2D materials are discussed in detail in Chapters 4–6, respectively. Finally, considering technological advancements, a comprehensive summary is provided in Chapter 7 along with future challenges and suggestions for developing efficient EMI shielding materials. It is intended that this book will assist scientists, researchers, and engineers in designing novel EMI shielding materials with customized structures and properties.

The chief editor and author, Chong Min Koo, would like to thank all the contributors to this handbook. It is worthy to note that this book is a product of long-standing research findings of the research group, Functional Polymer Hybrid Laboratory (FPHL) at Korea Institute of Science and Technology. FPHL has been working hard to develop the structure-designed materials for EMI shielding and various electronic applications since 2007. All the group members and alumni, including Dr. Soon Man Hong, Dr. Seung Sang Hwang, Dr. Kyung-Youl Baek, Hyunchul Park, Dr. Sangho Cho, Dr. Seon Joon Kim, Dr. Albert Lee, Dr. Faisal Shahzad, Dr. Pradeep Sambyal, Aamir Iqbal, Junpyo Hong, Dr. Jang-Woo Lee, Dr. Jin Hong Lee, Dr. Santosh Yadav, Dr. Richao Zhang, Dr. Hyesung Cho, Dr. Tae Hee Han, Dr. Bum Ki Baek, Dr. Seunggun Yu, Dr. Seung Hwan Lee, Dr. Won Jun Na, Dr. Pradip Kumar, Dr. Tae Yun Ko, Dr. Mirhani Seyyedalireza, Dr. Taegon Oh, Dr. Taehoon Kwon, Dr. Pradip Kumar, Dr. Richao Zhang, Kwang Ho Kim, Hang-Kyu Cho, Hyeong Tae Kim, Bori Kim, Myeong Hee Kim, Heela Kwak, Yong Jin Kwon, Kyongho Min, Minho Kim, Il Jin Kim, Ji Wook Shin, Yun Ho La, Seok Jin Noh, Byeori Ok, Younduk Park, Ji Young Jung, Hyerim Kim, Daesin Kim, Ki Hwan Koh, Yong Soo Cho, Han-Na Kim, Ari Chae, Sehyun Doo, Hwan Gyu Lee, Juyun Lee, Seung Jun Lee, Suung Chae, Soobin Kim, directly or indirectly, contributed to this research journey. Chong Min Koo would also like to express great gratitude and love to his wife, Professor Nogin Chung and his beloved son, Hasong Koo, for their great support. Enjoy.

November 2020
Seoul, Korea

Chong Min Koo
Pradeep Sambyal
Aamir Iqbal
Faisal Shahzad
Junpyo Hong

1

Electromagnetic Interference and Shielding

1.1 Introduction

Telecommunication devices and microelectronics inadvertently receive, generate, and/or propagate electromagnetic waves (EMWs) in a wide frequency range. Technological advancements toward smaller and smarter electronic devices have inevitably been accompanied by increased electromagnetic interference (EMI). EMI is a type of cross-talk between different devices and circuits operating in close proximity. This disrupting phenomenon often results in critical component malfunctioning, device underperformance, data loss, and incorrect signal interpretation [1–5]. More broadly, EMI is a serious concern for the aviation industry, including military jets, warships, and other strategic components, and can place a country's security at risk [6, 7]. The smallest error resulting from incorrect signal generation or interpretation could have serious consequences because of the false triggering of ammunition. Additionally, the increasing density of EMWs has a significant impact on human health [8, 9].

In the era of emerging fifth generation (5G) technology, the number of electronic devices and gadgets will increase drastically, resulting in increased exposure to wireless fidelity (Wi-Fi) and the internet of things (IoTs), as well as spreading electromagnetic terrorism and offensive information warfare [10]. Therefore, it is necessary to protect humans and electronic devices from the detrimental effects of EMI by providing suitable shielding. Thus, advanced EMI shielding materials should be developed to meet the challenges of advanced technologies.

The energy of EMWs can be attenuated by reflection, absorption, and multiple reflection mechanisms [11–13]. The primary mechanism involves the reflection of EMWs that strike the surface of a shielding material [2]. Highly conductive materials with excess mobile charge carriers (holes and/or electrons) show strong reflection upon interaction with electromagnetic radiation. Metals (e.g. Ag, Cu, and Al), which are the most conductive materials, show excellent EMI shielding properties through reflection, and have been used for decades in commercial appliances [14–16].

The second mechanism involves the absorption of EMWs within a shielding material. When EMWs propagate in a shielding material, their intensity is exponentially attenuated with the thickness. For efficient absorption, the electrical conductivity,

Two-Dimensional Materials for Electromagnetic Shielding, First Edition.
Chong Min Koo, Pradeep Sambyal, Aamir Iqbal, Faisal Shahzad, and Junpyo Hong.

dielectric permittivity, and magnetic permeability of the shielding material play a vital role in attenuating the energy of EMWs through Ohmic loss, dielectric loss, and magnetic loss, respectively [17–20].

The third mechanism involves multiple reflections that occur because of either thin thickness or the presence of multiple interfaces within the material. These types of multiple reflections are different and hence perform differently. When the thickness of a material is smaller than the skin depth, multiple reflections will decrease the total EMI shielding effectiveness (SE) value [12, 21], whereas this effect is negligible when the thickness of the material is larger than the skin depth. The skin depth is the thickness of the material at which the intensity of the incident EMWs falls by a value of $1/e$. In contrast, multiple reflections caused by the presence of multiple internal interfaces have a positive impact on the EMI SE value, as the internal scattering of EMWs from the internal interfaces will increase absorption and hence the EMI shielding ability of the material. A comprehensive description of each mechanism along with the influencing and controlling parameters is provided in Chapter 2.

1.2 Electromagnetic Field Sources and Impact on Human Beings

The innovation race is backed by advances in technology, which are marvelous in facilitating human life. Global technologies, including artificial intelligence (AI), automation, and IoT, provide services that affect every aspect of life. However, this emerging technology can also place human life at risk. The operating frequencies in the electromagnetic spectrum can be divided into two categories depending on potential hazards: ionizing and nonionizing (Figure 1.1). As the risks of high-frequency ionizing electromagnetic (EM) radiation, including X-rays and gamma cosmic rays, are well known, safety protocols have been developed to avoid any kind of exposure. Unfortunately, nonionizing low-frequency EMWs are a silent killer for humans and operating systems if left unattended. Currently, the most commonly used electronic device that operates in this frequency range is the cellular phone, the use of which has grown dramatically since the start of commercial services in 1983 [22]. In the first decade, the number of cellular users in the United States only reached ~2 million. However, there are now nearly 3 billion global users, and this number is expected to hit 6.1 billion in 2020 [22]. Although cellular phones and handheld devices transmit very low powers of 0.6 W while operating at 850 MHz to 3.5 GHz or even higher frequency, their frequent use in close proximity to the body is associated with severe health outcomes. Therefore, proper shielding of the field is required for a healthy work environment.

We are always surrounded by betraying electromagnetic fields. Generally, an alternating electric field generates a magnetic field around it, and vice versa. To apply appropriate shielding, it is essential to know about the sources that generate electromagnetic fields. These sources can be either natural or artificial (manmade).

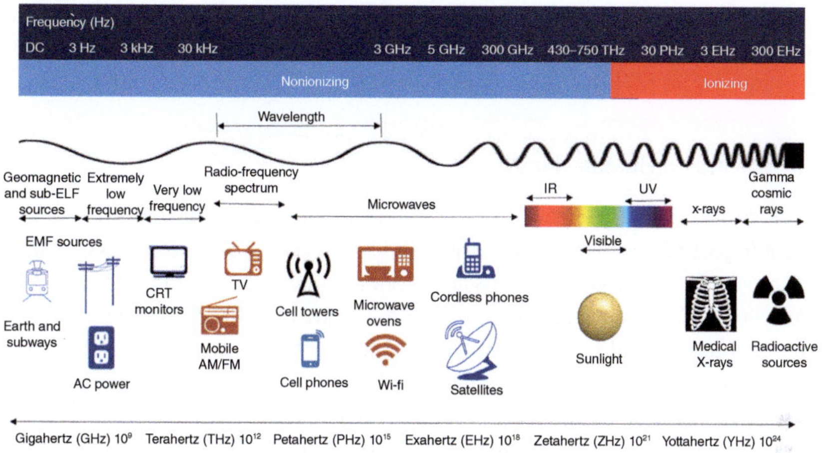

Figure 1.1 Frequency and wavelength ranges in the electromagnetic spectrum, and devices that emit radiation at certain frequencies. Source: Chong Min Koo.

1.2.1 Natural Sources

Generally, the Sun is the biggest source of electromagnetic radiation. Solar energy, which is the energy that reaches Earth from the Sun, consists of electromagnetic radiation at all wavelengths. Almost half of the radiation (49%) is in the low-frequency (longer wavelength) region, 44% in the visible region, which can be seen by the naked eye, and the remaining 7% in the high-frequency (shorter wavelength) ultraviolet region. This ionizing radiation is harmful to human skin and eyes, as well as the brain and various cells. In addition, the Earth, thunderstorms, and lightning also create strong electromagnetic fields. Earth's own magnetic field is strong enough to rotate a compass needle in the north–south direction. However, the generated geomagnetic field has a low frequency (1–300 Hz) with a power of 60 μT at the North/South poles and 30 μT at the equator [23]. Thunderstorms and lightning also produce electric fields by building up electrically charged particles in the atmosphere.

1.2.2 Artificial (Manmade) Sources

In the era of technology, all electrical or electronic devices used in everyday life (e.g. mini power sockets) generate electromagnetic radiation or require electromagnetic fields to function. Power lines and electronic equipment, including vacuum cleaners and hair dryers, are sources of low-frequency radiation with a power of approximately 17.44–164.75 μT when measured at a distance of 5 cm [23]. In contrast, radio stations, television antennas, mobile electronic devices, microwave ovens, and most importantly medical equipment generate high-frequency EMWs (100 kHz to 300 GHz). Typically, the radiation emitted from manmade sources is polarized, as it is produced by electromagnetic oscillation circuits. As polarized electromagnetic radiation can penetrate the human body, it is used in biomedical applications. However, exposure to such radiation above a certain limit is potentially deleterious to cells and tissues [24].

1.2.3 Effects on Human Health

We are surrounded by electromagnetic fields that have inevitably become a formidable part of our life. Initially, these fields were considered too weak to affect the biomolecular systems of humans and not strong enough to influence physiological functions. However, exposure to electromagnetic radiation can have a dreadful impact on human health. Some individuals are very sensitive to electromagnetic fields and show the symptoms of health disorders under mild exposure. This adverse phenomenon, called electromagnetic hypersensitivity (EHS), affects 1–3% of the global population, as reported by the WHO [25]. The observed symptoms include headaches, sleep disorders, dizziness, depression, palpitations, hot flushes, sweating, tinnitus, fatigue, limb pain, back pain, heart disease, tremors, nervousness, nausea, skin rashes, weakness, loss of appetite, and breathing difficulties [26]. Furthermore, sleep disorders and deficiencies have been observed in inhabitants close to electromagnetic radiation emitters, although several studies have also found otherwise. A case study found psychological symptoms such as depression, unmanageable emotions, and suicide among residents exposed to the 50 Hz chronic frequency of high-voltage power stations and highlighted an increased suicide (attempt) rate, most likely because of depression [27].

The ear is the first organ to be exposed to the electromagnetic radiation emitted by cellular phones, and a study reported that exposure to electromagnetic radiation of 50 Hz at an intensity of 4.45 pT may cause adverse auditory effects in humans [28]. The cochlear outer hair cells are vulnerable to injuries from exogenous and endogenous agents and electromagnetic radiation. To numerically analyze the specific absorption rate (SAR) along with heat transfer in the human eye, Wessapan and Rattanadecho exposed a human eye model to an electromagnetic field of 900–1800 MHz [29]. This study revealed heat and mass transfer phenomena in the eye under exposure to electromagnetic fields, with the highest SAR observed in the cornea. Lower frequencies affected the anterior chamber, whereas higher frequencies increased the temperature in the vitreous. The exposure time also strongly influenced the temperature increase in the human eye. These results demonstrate the fatal effects of electromagnetic fields on the visual performance of the human eye.

Epidemiological investigations have shown that electromagnetic radiation is carcinogenic to humans, with long exposure causing tumor development. These effects are more apparent in newborns. Studies on the effects of electromagnetic radiation on the nervous tissues have found that high-intensity radio frequency (RF) exposure affects the central nervous systems, brain chemistry, and blood–brain linkages in animals. Lowered concentrations of dopamine, epinephrine, and norepinephrine in the brain are associated with electromagnetic radiation. Ghione et al. experimentally observed altered nociception and cardiovascular abnormalities in a human head

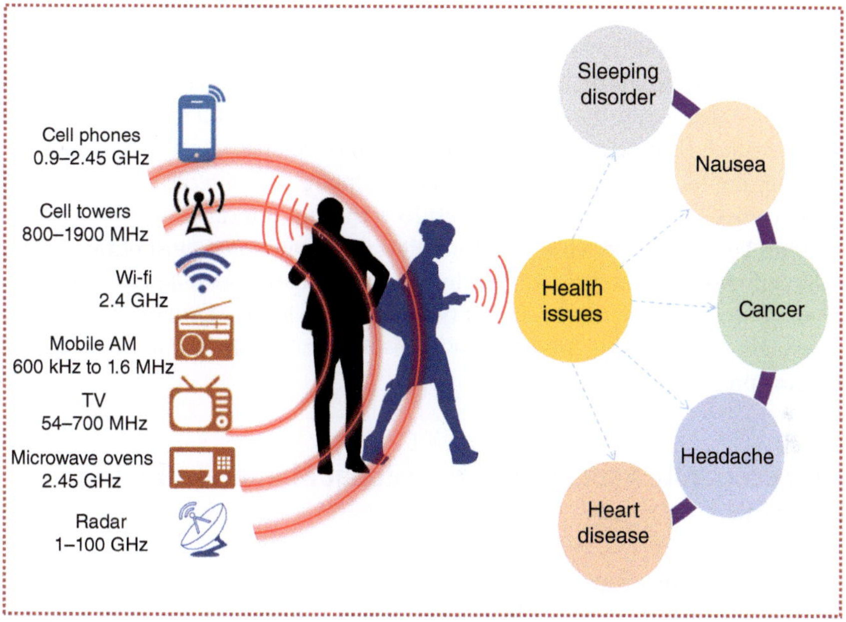

Figure 1.2 EMW sources and their potential harmful effects on humans. Source: Chong Min Koo.

exposed to electromagnetic radiation of 37 Hz with a flux density of 80 μT [30]. When placed close to the location of the heart in the chest, betrayed electromagnetic fields could also interfere with implanted pacemakers by altering heartbeats. Importantly, decreased male fertility is also linked with exposure to electromagnetic fields. Cellular phones have become an integral part of everyday life and are carried in pockets, very close to the reproductive system. Therefore, it is particularly important to investigate the potential effects of operating frequency and power. In a study on rats, various histopathological alterations, such as necrosis, focal tubular atrophy, and seminiferous epithelial erosion, were recorded under low-frequency exposure at 50 Hz; however, the serum testosterone level was barely affected [31]. Figure 1.2 summarizes various EMW sources along with their potential detrimental effects on human health.

Therefore, a practical way to safeguard humans is to follow the ALARA (as low as reasonably achievable) principle by maintaining a safe distance and minimizing exposure or to take other precautionary steps such as efficient shielding to attenuate the energy of EMWs. As maintaining a safe distance is practically impossible, the shielding strategy has emerged as a potential solution.

1.3 EMI Hazards for Data Security

EMI is a phenomenon in which the electromagnetic radiation from one electronic device disrupts other nearby electronic circuits via conduction or radiation transfer.

A severe EMI scenario can adversely affect the entire electronic system, causing device malfunction or system failure. This effect is quite strong in medical equipment, where electromagnetic radiation is used for biomedical applications and imaging. Consequently, the figures and/or results are affected if effective shielding is not provided. Low-frequency EMI caused by power sources can also have harmful effects on the hardware, resulting in data corruption or complete reformatting of the hard disk in severe cases. As a result, retrieving information from wireless terminals becomes impossible. Although EMI is unintentional most of the time, it can be misused in electronic warfare in the form of data loss, data hacking, radio jamming, security breaches, and other types of disturbances. Therefore, the defense department of a country strictly deals with disastrous EMI by following special shielding protocols.

1.4 Economic Aspects and the Global Market for EMI Shielding

From an application point of view, the market for EMI shielding materials can be categorized into different areas, including electronics, defense, aerospace, automotive, telecommunications, and medical appliances. Electronics hold a substantial market share because advanced compact and fast devices use efficient circuits that operate at higher frequencies. The second highest market share belongs to the defense sector, with applications including sophisticated satellites, radar equipment, and spacecraft. Countries are continuously increasing their defense budgets for the research and development of new arms and weapons, thus increasing global defense expenditures and contributing to a larger electromagnetic absorption and shielding market. In 2015, global military expenditures were approximately US$ 1.5 trillion, led by the United States, China, and Saudi Arabia (see Figure 1.3a for other countries). The aerospace industry is also growing with various ongoing space exploration missions. Furthermore, nearly 30 000 new passenger aircraft will be added to those currently in service over the next 20 years; hence, EMI shielding materials will be in high demand. Similarly, telecommunication devices, AI, IoT, and the automotive industry are expected to thrive in the coming years, leading to an ever-increasing demand for EMI shielding materials.

The global EMI shielding market has a projected size of US$ 6.8 billion in 2020 and is expected to reach US$ 9.2 billion by 2025, with a compound annual growth rate of 6.3% over the next five years (Figure 1.3b) [32]. Thus, there is a huge demand for the research and development of novel and efficient materials to meet stringent EMI shielding requirements.

1.5 Electromagnetic Compatibility Regulations and Standards

Over the past few decades, the increase in electrical and electronic systems with their accompanying electromagnetic pollution has necessitated the creation and

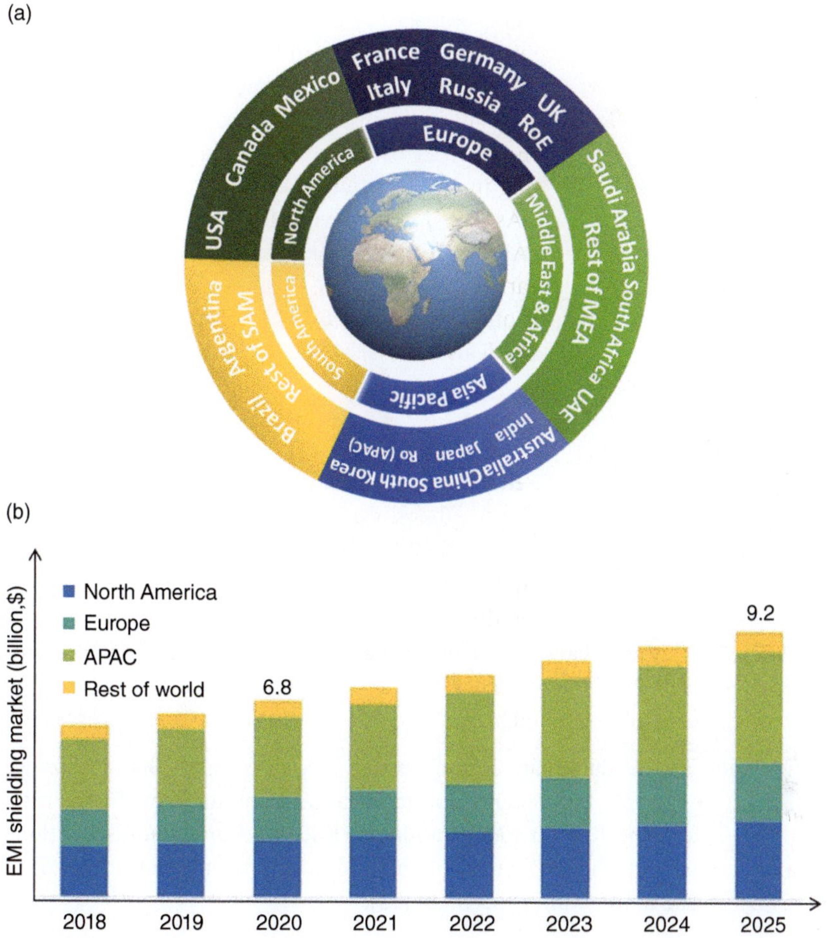

Figure 1.3 (a) Global market shares in the field of EMI shielding and (b) global market over the next five years. Source: https://www.marketsandmarkets.com/Market-Reports/emi-shielding-market-105681800.html.

implementation of electromagnetic compatibility (EMC) standards around the globe (Figure 1.4). Historically, during World War II, EMC became very important in the defense sector, as mission success was highly reliant on electronic communication systems, which is still the case today. In the civil sector, the extensive use of radio facilities, modern gadgets, telecommunications, and electric systems has spread EMI pollution. Therefore, a set of standards is required so that the products or systems applied in the civil, defense, and industry sectors are immune to external EMI pollution and do not generate high electromagnetic radiation.

Compiling all existing standards and creating a set of documents as a recommendation for EMC is tedious, as every nation has its own set of standards for instruments, procedures, and limits. The most renowned international organizations are the International Standard Organization (ISO) and the International Electrotechnical Commission (IEC). Alongside these two agencies, many other

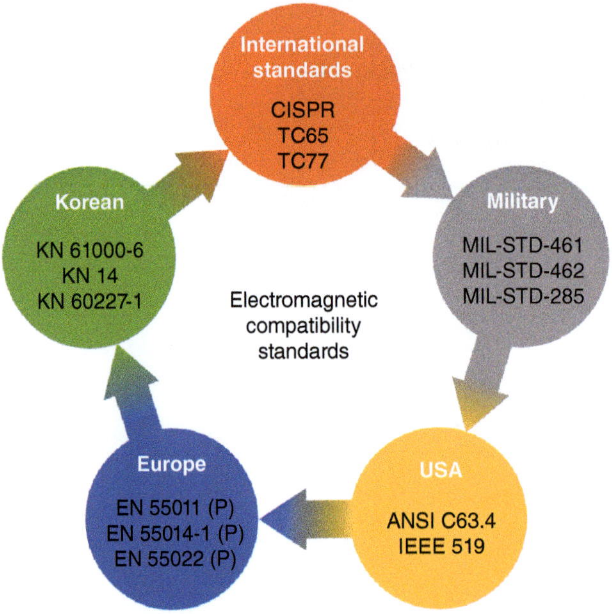

Figure 1.4 Summary of global standards and limitations for the safe use of electrical and electronic equipment. Source: Chong Min Koo.

agencies, such as the European Committee for Electrotechnical Standardization (CENELEC) in Europe, the Federal Communications Commission (FCC) in the United States, and the Korean EMC standards (KSC) in Korea, publish national standards for use in the civilian segment and military or defense standards (MIL-STD) for use in the military segment [33].

1.5.1 International Standards

The IEC, which is the organization that monitors EMC standards and testing procedures, comprises three technical committees on EMC, namely, CISPR (International Special Committee on Radio Interference), TC65 (related to the immunity standards), and TC77 (related to EMC in electrical systems and networks). CISPR compiles their findings related to EMC issues in the form of standards labeled CISPR10–CISPR23. TC65 provides standards related to immunity under designation 801. TC77 provides standards related to low-frequency phenomena in the power network, such as harmonics and flicker, under designations 555 and 1000, respectively. The CISPR22 standards are the basic foundation of the national standards of several countries for emission from IT or communication systems. Owing to variations in the prerequisites of various countries, the standards suggested by CISPR cannot be implemented as law. However, in accordance with the will of each nation, these standards can be implemented or modified as national standards [34].

1.5.2 FCC Standards (United States)

In the United States, radio and wired communications and interference are regulated by the FCC. Any equipment that is imported to or sold in the United States must fulfill the standards laid out by the FCC. The FCC rules and regulations deal with EMC standards, measurement methods American National Standards Institute (ANSI C63.4), equipment or product approval processes, and marking requirements. Three sections deal with EMC, namely, Part 15 (for radio-frequency systems), Part 18 (for industrial, scientific, and medical equipment), and Part 68 (for equipment connected to the telephone network). The standard regulations divide devices or equipment into two categories: Class A and Class B. Class A includes systems used in the commercial, industry, and business sectors, whereas Class B includes domestic systems. The regulation limits for Class B (approximately 10 dB) are stricter than those for Class A. All systems with digital circuits and a clock frequency above 10 kHz fall within the FCC regulations. The conducted interference limits are listed in the range of 450 kHz to 30 Hz, and the present limit controls the interference current in power leads. The leading organization for EMC standardization in the United States is the ANSI, to which various other institutions also contribute, such as the IEEE EMC Society [34, 35].

1.5.3 European Standards

The European Standard Committee (ESC) monitors the regulation and implantation of suitable standards using CENELEC. These standards, including basic standards (test methods) and product standards (specific types of equipment), classify systems or equipment into broad classes, namely, class 1 (residential and light industrial) and class 2 (heavy industry). The mandate of 3 May 1989, related to EMC prerequisites, was amended for a transitional period and came into effect on 1 January 1996. All the member nations have to sanction the legislation to implement these orders. The standard orders are widely inclusive, accounting for the emission and susceptibility of various hardware or systems and imposing duties on the manufacturers, irrespective of the existence of suitable guidelines. A trade agreement was signed in 1997 between the FCC and CENELEC, through which any EMC test conducted in either the United States or the EU is acknowledged by both parties. Thus, trade barriers were demolished, and both European and US manufacturing companies and testing laboratories can export and import equipment to each other [34].

1.5.4 Korean Standards

Korean EMC and safety standards are implanted and maintained by the Radio Research Agency (RRA) and the Korean Agency for Technology and Standards (KATS). The standards are classified into two categories: EMC (Radio Wave Law) and safety (Electric Appliances Safety Control Act). The first EMC policy was converted into the Radio Wave Act in 1989, and the electromagnetic susceptibility (immunity) criteria were enforced in 2000. The Korean EMC standards (KS) are identical to the IEC standards. For example, the KN 61000-6 family of standards is identical to the IEC 61000 standards. The KN 14 standard, which deals with household and electrical systems, is analogous to the CISPR 14-1 standard. Moreover, all products that meet Korean EMC and electrical safety requirements are engraved with a KC mark before being introduced to the Korean market. In July 2011, the EU and Korea established a system that facilitates the certification processes for electrical and electronic systems, thus increasing market accessibility.

1.5.5 Military or Defense Standards

The most important group of EMC standards is the military standards. In modern warfare, electric and electronic communications are used to set up integrated control commands among all the forces (land, air, and sea). Hostile warfare requires compact and confidential systems. Moreover, the high use of electromagnetic energy for jamming has generated critical EMC systems and EMI complications. Any electrical system designed and developed for defense purposes must fulfill the MIL-STD-461 standard, which gives a limit range for the susceptibility of equipment to conduct and radiate EMI. The MIL-STD-462 companion standard provides testing procedures. Moreover, the MIL-STD-461 standards have been acknowledged and implemented outside the United States by various defense establishments and a few nonmilitary organizations. Military equipment should be able to withstand radiated and conducted RF without malfunctioning. The military standards are stricter than the civilian standards (FCC, ESC, and CISPR22) as they deal with both susceptibility and emission and cover a frequency range of 30–40 GHz. Strict military standards are needed for radiated emission because of the small size of the working space, where electronic systems are kept in close proximity in fighter jets, tanks, naval ships, etc. In contrast, these systems can be more widely distributed in industrial installations [34, 35].

1.6 Materials for EMI Shielding

The rise of electronic and electrical devices started after the development of the electromagnetic theory in the nineteenth century. The understanding of the electromagnetic theory has led to the use of EMWs for wireless communication, information sharing and broadcasting, foreign object detection by security systems, and medical applications and imaging. In general, every device currently working on a power source receives or generates undesirable EMWs. Devices operating in close proximity can interfere with each other, resulting in malfunctioning or failure and can also affect human health. Therefore, in this advanced technology era, as we cannot reduce the use of electronic equipment, huge efforts are being made to develop efficient materials to mitigate or absorb the energy of undesirable EMWs.

As mentioned previously, reflection is the primary mechanism of EMI shielding, as a major portion of the incident EMWs is reflected after striking the surface of a shielding material. In this regard, highly conductive metals (e.g. Ag, Cu, and Al) in the form of foils or shrouds have been commonly used for decades as potential EMI shielding materials [2]. Owing to their high electrical conductivities of 10^5–10^6 S cm^{-1}, these metals have abundant free electrons that can interact with the incident EMWs and reflect them back into space. The excellent thermal conductivities of these metals (200–500 W m^{-1} K^{-1}) also expand their applications. However, the shielding performance of highly conductive metals is outweighed by various drawbacks such as high density, high cost, poor corrosion resistance, and difficulties in processing at smaller thicknesses. These factors limit the applications of metals in the aircraft and aerospace industry, where lightweight materials are a priority. Moreover, the reflection mechanism generates secondary pollution when the reflected EMWs interact with surrounding circuits. Therefore, in advanced portable and smart devices, efficient absorbing materials are being extensively explored.

Although electrical conductivity is a critical parameter for determining the performance of an EMI shielding material, it is not the only factor. Ferromagnetic materials (paramagnetic materials) show strong absorption behavior owing to spontaneous magnetization below the Curie temperature [36]. Under an applied magnetic field, the spin moments in the domains (a microscopically large homogeneous region) are aligned parallel to the magnetic field, resulting in the storage of the energy associated with electromagnetic radiation. Thus, the magnetic permeability of a material is an important criterion for defining the ability of the material to absorb the energy of EMWs. As a result, ferrites have become of significant academic and industrial interest to overcome the disadvantages of highly conductive metals.

Conductive polymer nanocomposites with metal nanoparticles and ferrite inclusions have excellent capabilities for EMI shielding. In particular, cost-effective and easily processable composites with multiphase heterogeneous structures hosting carbon nanotubes (CNTs) or carbon black (CB) have received considerable attention.

Currently, two-dimensional (2D) nanomaterials are at the forefront of ongoing research owing to their unique chemical, mechanical, electrical, and optoelectronic properties [37–41]. Since the discovery of graphene in 2004, research on 2D materials has proceeded at a far higher rate than ever before [42]. Nevertheless, this field is still emerging, with new materials being discovered every year and many more anticipated in the future. The range of 2D materials includes graphene [5, 43–45], black phosphorous (BP), [46] hexagonal boron nitride (h-BN) [47], transition metal dichalcogenides (TMDCs; e.g. MoS_2, WS_2, TaS_2, $MoSe_2$, and WSe_2), [48–50] and the very new yet gigantic family of transition metal carbides/nitrides/carbonitrides (MXenes; e.g. $Ti_3C_2T_x$, Ti_3CNT_x, and Ti_2CT_x) [51–54]. The development of novel 2D materials beyond graphene has revealed a wide range of electrical conductivities (0.004–15 000 S cm^{-1}, Table 1.1), which imparts extraordinary potential for diverse applications in almost all technological areas. Owing to their excellent electrical conductivity, processability, corrosion resistance, and low density, 2D materials are believed to be an alternative to highly conductive metals for EMI shielding. Monolayer graphene shows outstanding EMI shielding potential because of its very high electrical conductivity (5.25×10^4 S cm^{-1}) in chemical vapor deposition

Table 1.1 Fundamental properties of conventional shielding materials and advanced 2D nanomaterials.

Material		Properties							
		Conductivity ($S\,cm^{-1}$)	Permeability (μ′/μ″)	Carrier density (cm^{-3})	Carrier mobility ($cm^2\,V^{-1}\,s^{-1}$)	Density ($g\,cm^{-3}$)	Thermal conductivity ($W\,m^{-1}\,K^{-1}$)	Electronic bandgap (eV)	References
Metals	Silver	6.305×10^5	—	5.86×10^{22}	9490	10.53	417–427	N/A	[55, 56]
	Copper	5.977×10^5	—	8.47×10^{22}	5770	8.9	386–400	N/A	[55, 56]
	Aluminum	3.538×10^5	—	18.1×10^{22}	2600	2.7	234	N/A	[55, 56]
Magnetic materials	Fe_2O_3	3.12×10^{-7}	1.2/0.8(4 GHz, 70 wt%/wax)	1.65×10^{19}	$\sim 10^{-2}$	5.25	—	2.0–2.4	[57–64]
	Fe_3O_4	$\sim 10^2$–10^3	1.4/0.9 (4 GHz, 70 wt%/wax)	4.04×10^{22}	—	5.17	2.7 (cross) 2.6 (in-plane)	2.3	[57, 58, 60, 63–70]
	Sendust	1.397	4/1.5 (1 GHz, 40 vol%/acryl)	—	—	2.4–3.5	1.83 (50 vol%/PP)	—	[71–73]
	Permalloy	0.8	9/8 (1 GHz, 70 vol%/PPS)	$7.9 \times 10^{11}\ cm^{-2}$	0.5×10^6	8.7	19	—	[74–76]
	MnZn ferrite	0.008 (71 vol%/PANI)	3/4 (1 GHz, 71 vol%/PANI)	—	—	4.8	4	2.19	[76–78]
	NiZn ferrite	1.35×10^{-1} (54 wt%/PANI)	1–2/7–8 (1 GHz)	—	—	5.2	7	1.91	[76, 79–82]
2DGraphene materials	CVD graphene	5.25×10^4	—	2.8×10^{13}	4.4×10^4	—	2500 ± 1000	0.01	[83–89]
	Graphene oxide	(Insulating)	—	—	0.25	2.2	776	1.7	[90–93]
	Reduced graphene oxide	7–205	—	1.3×10^{11}	16–262	2.1	1390	2.2–0.5	[94–101]

	S-doped graphene	21–311	—	2.3×10^{18}	60–270	2.25	N/A	0.1–0.2	[99, 102–104]
	N-doped graphene	12–316	—	2.5×10^{15} (N-doped) 6×10^{14} (B-doped)	350–550 (N-doped) 450–650 (B-doped)	~2	4–300 (N-doped) 190 (B-doped)	0.2 eV (N-doped) 0.6 eV (N,B-codoping)	[105–113]
MXenes	$Ti_3C_2T_x$	5000–15 000	—	$(2.6–3.2)\times10^{16}$ cm^{-2}	(0.6–1.25)	2.39	2.84	0.5–2	[114]
	Ti_3CNT_x	1125–2712	—	—	—	—	—	—	[4, 115]
	Ti_2CT_x	1600–2000	—	1×10^{22}	33.6×10^3	~2	11.91	0.88–1.15	[116–119]
	V_2CT_x	998–1560	—	—	—	—	—	—	[115]
	Nb_2CT_x	5	—	—	$\sim1\times10^6$	—	15	—	[115, 120, 121]
	Mo_2CT_x	0.017–4.3	—	—	—	—	48.4	—	[122]
	Sc_2CF_2	—	—	—	$(1.07–5.03)\times10^3$	—	178–472	1.03	[119, 123]
	$Sc_2C(OH)_2$	—	—	—	$(2.06–2.19)\times10^3$	—	107–173	0.45	[119, 123]
TMDCs	MoS_2	1000	—	2.6×10^{12} cm^{-2}	50–217	5.06	34.5–131	1.3 (indirect)	[124–131]
	WS_2	6.7	—	1.0×10^{17} cm^{-2} (thin film) $3.0\times10^{17}\sim1.6\times10^{18}$ cm^{-3} (nanotube)	0.2	7.5	32 (monolayer) 53 (bilayer)	1.35 (bulk) 2.05 (monolayer)	[131–137]
	TaS_2	680	—	4.5×10^{19} cm^{-3}	20	6.86	11.55–13.36	0.3	[138–141]
Others	CuS	870	—	2.09×10^{21} cm^{-3}	1–6	4.76	1.8	2.0	[142–145]
	h-BN	(Insulating)	—	—	2300	2.28	280	4.9–6.4	[146, 147]
	BP	0.004	—	2.6×10^{14} cm^{-2}	1000	2.69	110 (AC) 35 (ZZ)	0.3 (bulk) 2.0 (monolayer)	[148–156]

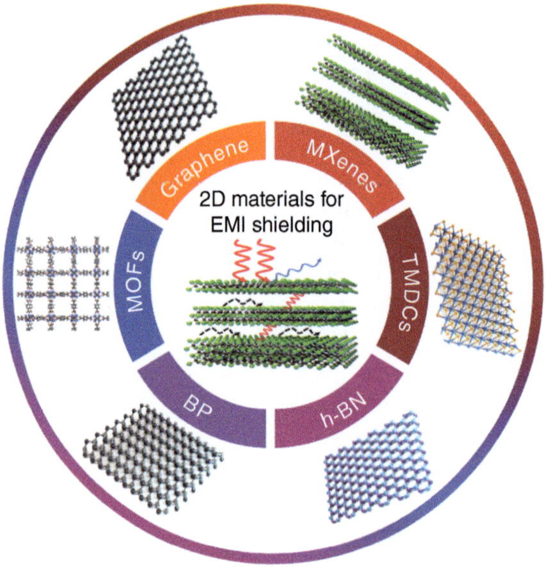

Figure 1.5 Atomic structures of 2D materials used for EMI shielding. The development of these novel 2D materials has provided a breakthrough in replacing conventional conductive metals, which suffer from high density and corrosion susceptibility. Source: Chong Min Koo.

(CVD)-grown nanometer thin films. At higher thicknesses, the electrical conductivity decreases because the defect concentration increases. Metallic conductivities of more than 15 000 S cm^{-1} have been achieved for MXenes at thicknesses of nanometers to tens of micrometers, and their polymeric composites show an ultralow percolation threshold with superior electrical conductivity. Owing to their unique advantages for fabricating highly processable liquid dispersions, inks, low-percolation polymer composites, and flexible and lightweight composites, 2D materials can satisfy the requirements for on-ground and space applications. TMDCs with 2D morphologies, including WS_2, TaS_2, and MoS_2, show high electrical conductivities of 6.7, 680, and 1000 S cm^{-1}, respectively. CuS, another 2D material, also exhibits a high electrical conductivity of 870 S cm^{-1}. The synergistic effect of efficient electrical conductivity and a multilayered morphology makes these 2D materials effective for practical EMI shielding applications. The fundamental properties of conventional and novel EMI shielding materials are summarized in Table 1.1, and these materials, especially 2D graphene, MXenes, TMDCs, BP, h-BN, and MOF (Figure 1.5), are discussed in detail in Chapters 4–6.

The current rise in 5G technology demands more efficient EMI shielding materials with ultralow densities and thicknesses. The area of EMI shielding materials is emerging parallel to technological advancements, and scientists are using a structural control approach to further enhance and tailor the properties of 2D materials according to the desired area of application. In this way, compact solid films and composites, porous foams and aerogels, and segregated structures have been developed for use as thin, strong, lightweight, and cost-effective shielding materials [11].

1.7 Summary

The rapid advancement of highly integrated electronic, telecom, defense, and medical devices has resulted in serious EMI, which can detrimentally impact the performance and lifetime of these devices as well as human health. Although metals and magnetic materials have long been used for EMI shielding, they are not a good choice for lightweight and portable technologies. Novel 2D materials, which possess high electrical conductivity, low density, high mechanical strength, and excellent EMI SE at a minimal thickness, can provide the same EMI shielding performance as metals while minimizing the drawbacks. The shielding and absorption of EMWs is pivotal in a wide variety of industries, including satellites, aerospace, and defense, which provide a huge global market for EMI shielding and absorbing materials. Therefore, efficient materials with the required set of fundamental properties are highly needed, and thus continue to be explored. This book provides insights into the potential of 2D materials for lightweight EMI shielding.

References

1 Simon, R.M. (1981). EMI shielding through conductive plastics. *Polymer-Plastics Technology and Engineering* 17 (1): 1–10.

2 Shahzad, F., Alhabeb, M., Hatter, C.B. et al. (2016). Electromagnetic interference shielding with 2D transition metal carbides (MXenes). *Science* 353 (6304): 1137–1140.

3 Chen, Z., Xu, C., Ma, C. et al. (2013). Lightweight and flexible graphene foam composites for high-performance electromagnetic interference shielding. *Advanced Materials* 25 (9): 1296–1300.

4 Iqbal, A., Shahzad, F., Hantanasirisakul, K. et al. (2020). Anomalous absorption of electromagnetic waves by 2D transition metal carbonitride Ti_3CNT_x (MXene). *Science* 369 (6502): 446–450.

5 Hong, S.K., Kim, K.Y., Kim, T.Y. et al. (2012). Electromagnetic interference shielding effectiveness of monolayer graphene. *Nanotechnology* 23 (45): 455704.

6 Lee, C., Lee, D., Jeong, C. et al. (2002). Electromagnetic interference shielding by using conductive polypyrrole and metal compound coated on fabrics. *Polymers for Advanced Technologies* 13 (8): 577–583.

7 Geetha, S., Satheesh Kumar, K., Rao, C.R. et al. (2009). EMI shielding: methods and materials—a review. *Journal of Applied Polymer Science* 112 (4): 2073–2086.

8 Nazir, A., Yu, H., Wang, L. et al. (2018). Recent progress in the modification of carbon materials and their application in composites for electromagnetic interference shielding. *Journal of Materials Science* 53 (12): 8699–8719.

9 Stam, R. and Yamaguchi-Sekino, S. (2018). Occupational exposure to electromagnetic fields from medical sources. *Industrial Health* 56 (2): 96–105.

10 Gurevich, V. (2005). The hazards of electromagnetic terrorism. *Public Utilities Fortnightly* 143: 84–86.

11 Iqbal, A., Sambyal, P., and Koo, C.M. (2020). 2D MXenes for electromagnetic shielding: a review. *Advanced Functional Materials* 30: 2000883.

12 Yun, T., Kim, H., Iqbal, A. et al. (2020). Electromagnetic shielding of monolayer MXene assemblies. *Advanced Materials* 32 (9): 1906769.

13 Lee, S.H., Yu, S., Shahzad, F. et al. (2017). Density-tunable lightweight polymer composites with dual-functional ability of efficient EMI shielding and heat dissipation. *Nanoscale* 9 (36): 13432–13440.

14 Choi, Y.-S., Yoo, Y.-H., Kim, J.-G. et al. (2006). A comparison of the corrosion resistance of Cu–Ni–stainless steel multilayers used for EMI shielding. *Surface and Coatings Technology* 201 (6): 3775–3782.

15 De Temmerman, L. (1992). New metallized materials for EMI/RFI shielding. *Journal of Coated Fabrics* 21 (3): 191–198.

16 Luo, X. and Chung, D. (1999). Electromagnetic interference shielding using continuous carbon-fiber carbon-matrix and polymer-matrix composites. *Composites Part B Engineering* 30 (3): 227–231.

17 Kim, S.-S., Kim, S.-T., Yoon, Y.-C. et al. (2005). Magnetic, dielectric, and microwave absorbing properties of iron particles dispersed in rubber matrix in gigahertz frequencies. *Journal of Applied Physics* 97 (10): 10F905.

18 Bhadra, S., Singha, N.K., and Khastgir, D. (2009). Dielectric properties and EMI shielding efficiency of polyaniline and ethylene 1-octene based semi-conducting composites. *Current Applied Physics* 9 (2): 396–403.

19 Yin, X., Xue, Y., Zhang, L. et al. (2012). Dielectric, electromagnetic absorption and interference shielding properties of porous yttria-stabilized zirconia/silicon carbide composites. *Ceramics International* 38 (3): 2421–2427.

20 Chen, Y., Zhang, H.-B., Huang, Y. et al. (2015). Magnetic and electrically conductive epoxy/graphene/carbonyl iron nanocomposites for efficient electromagnetic interference shielding. *Composites Science and Technology* 118: 178–185.

21 Bosman, H., Lau, Y.Y., and Gilgenbach, R.M. (2003). Microwave absorption on a thin film. *Applied Physics Letters* 82 (9): 1353–1355.

22 Bren, S.A. (1996). Historical introduction to EMF health effects. *IEEE Engineering in Medicine and Biology Magazine* 15 (4): 24–30.

23 Jaroszewski, M., Thomas, S., and Rane, A.V. (2018). *Advanced Materials for Electromagnetic Shielding: Fundamentals, Properties, and Applications*. Wiley.

24 Singh, M. (2014). *Introduction to Biomedical Instrumentation*. PHI Learning Pvt. Ltd.

25 Hansson Mild, K.; Repacholi, M.; Van Deventer, E.; Ravazzani, P. et al. (2004). Electromagnetic hypersensitivity: proceedings. *International Workshop on EMF Hypersensitivity*, Prague, Czech Republic.

26 Singh, S. and Kapoor, N. (2014). Health implications of electromagnetic fields, mechanisms of action, and research needs. *Advances in Biology* 2014: 198609.

27 Van Wijngaarden, E., Savitz, D.A., Kleckner, R.C. et al. (2000). Exposure to electromagnetic fields and suicide among electric utility workers: a nested case-control study. *Occupational and Environmental Medicine* 57 (4): 258–263.

28 Budak, B., Budak, G.G., Öztürk, G.G. et al. (2009). Effects of extremely low frequency electromagnetic fields on distortion product otoacoustic emissions in rabbits. *Auris Nasus Larynx* 36 (3): 255–262.

29 Wessapan, T. and Rattanadecho, P. (2013). Specific absorption rate and temperature increase in the human eye due to electromagnetic fields exposure at different frequencies. *International Journal of Heat and Mass Transfer* 64: 426–435.

30 Ghione, S., Del Seppia, C., Mezzasalma, L. et al. (2004). Human head exposure to a 37 Hz electromagnetic field: effects on blood pressure, somatosensory perception, and related parameters. *Bioelectromagnetics* 25 (3): 167–175.

31 Erpek, S., Bilgin, M.D., Dikicioglu, E. et al. (2007). The effects of low frequency electric field in rat testis. *Revue de Médecine Vétérinaire* 158 (4): 206–212.

32 EMI shielding market by material (EMI shielding tapes & laminates, conductive coatings & paints, metal shielding, conductive polymers, EMI filters), method (radiation, conduction), industry, and region – global forecast to 2025. https://www.marketsandmarkets.com/Market-Reports/emi-shielding-market-105681800.html#:~:text=%5B143%20Pages%20Report%5D%20The%20globalwww.marketsandmarkets.com (accessed 7 January 2021).

33 Celozzi, S., Araneo, R., and Lovat, G. (2008). *Electromagnetic Shielding*. Wiley.

34 Tong, X.C. (2016). *Advanced Materials and Design for Electromagnetic Interference Shielding*. CRC Press.

35 Ott, H.W. (1976). *Noise Reduction Techniques in Electronic Systems*. Wiley.

36 Sasaki, T., Sonoda, S., Yamamoto, Y. et al. (2002). Magnetic and transport characteristics on high curie temperature ferromagnet of Mn-doped GaN. *Journal of Applied Physics* 91 (10): 7911–7913.

37 Zhang, H. (2015). Ultrathin two-dimensional nanomaterials. *ACS Nano* 9 (10): 9451–9469.

38 Khan, K., Tareen, A.K., Aslam, M. et al. (2020). Recent developments in emerging two-dimensional materials and their applications. *Journal of Materials Chemistry C* 8 (2): 387–440.

39 Stanford, M.G., Rack, P.D., and Jariwala, D. (2018). Emerging nanofabrication and quantum confinement techniques for 2D materials beyond graphene. *NPJ 2D Materials and Applications* 2 (1): 20.

40 Zavabeti, A., Jannat, A., Zhong, L. et al. (2020). Two-dimensional materials in large-areas: synthesis, properties and applications. *Nano-Micro Letters* 12 (1): 66.

41 Hantanasirisakul, K. and Gogotsi, Y. (2018). Electronic and optical properties of 2D transition metal carbides and nitrides (MXenes). *Advanced Materials* 30 (52): 1804779.

42 Novoselov, K.S., Geim, A.K., Morozov, S.V. et al. (2004). Electric field effect in atomically thin carbon films. *Science* 306 (5696): 666–669.

43 Jia, L.-C., Yan, D.-X., Liu, X. et al. (2018). Highly efficient and reliable transparent electromagnetic interference shielding film. *ACS Applied Materials & Interfaces* 10 (14): 11941–11949.

44 Erdogan, N., Erden, F., Astarlioglu, A.T. et al. (2020). ITO/Au/ITO multilayer thin films on transparent polycarbonate with enhanced EMI shielding properties. *Current Applied Physics* 20 (4): 489–497.

45 Wang, H., Lu, Z., Liu, Y. et al. (2017). Double-layer interlaced nested multi-ring array metallic mesh for high-performance transparent electromagnetic interference shielding. *Optics Letters* 42 (8): 1620–1623.

46 Hanlon, D., Backes, C., Doherty, E. et al. (2015). Liquid exfoliation of solvent-stabilized few-layer black phosphorus for applications beyond electronics. *Nature Communications* 6 (1): 1–11.

47 Lin, Y., Williams, T.V., and Connell, J.W. (2010). Soluble, exfoliated hexagonal boron nitride nanosheets. *The Journal of Physical Chemistry Letters* 1 (1): 277–283.

48 Chhowalla, M., Shin, H.S., Eda, G. et al. (2013). The chemistry of two-dimensional layered transition metal dichalcogenide nanosheets. *Nature Chemistry* 5 (4): 263–275.

49 Huang, X., Zeng, Z., and Zhang, H. (2013). Metal dichalcogenide nanosheets: preparation, properties and applications. *Chemical Society Reviews* 42 (5): 1934–1946.

50 Chhowalla, M., Liu, Z., and Zhang, H. (2015). Two-dimensional transition metal dichalcogenide (TMD) nanosheets. *Chemical Society Reviews* 44 (9): 2584–2586.

51 Naguib, M., Kurtoglu, M., Presser, V. et al. (2011). Two-dimensional nanocrystals produced by exfoliation of Ti_3AlC_2. *Advanced Materials* 23 (37): 4248–4253.

52 Naguib, M., Mashtalir, O., Carle, J. et al. (2012). Two-dimensional transition metal carbides. *ACS Nano* 6 (2): 1322–1331.

53 Anasori, B., Xie, Y., Beidaghi, M. et al. (2015). Two-dimensional, ordered, double transition metals carbides (MXenes). *ACS Nano* 9 (10): 9507–9516.

54 Anasori, B. and Gogotsi, Û.G. (2019). *2D Metal Carbides and Nitrides (MXenes)*. Springer.

55 Huang, Y., Kormakov, S., He, X. et al. (2019). Conductive polymer composites from renewable resources: an overview of preparation, properties, and applications. *Polymers* 11 (2): 187.

56 Palenskis, V. (2013). Drift mobility, diffusion coefficient of randomly moving charge carriers in metals and other materials with degenerated electron gas. *World Journal of Condensed Matter Physics* 3: 73–81.

57 Meng, Y., Liu, X.-W., Huo, C.-F. et al. (2016). When density functional approximations meet iron oxides. *Journal of Chemical Theory and Computation* 12 (10): 5132–5144.

58 Sivakumar, G. (2017). Synthesis of γ-Fe_2O_3, Fe_3O_4 and copper doped Fe_3O_4 nanoparticles by sonochemical method. *Sains Malaysiana* 46 (10): 1935–1942.

59 Gao, Y., Wu, D., Wang, T. et al. (2016). One-step solvothermal synthesis of quasi-hexagonal Fe_2O_3 nanoplates/graphene composite as high performance electrode material for supercapacitor. *Electrochimica Acta* 191: 275–283.

60 Wu, H., Wu, G., and Wang, L. (2015). Peculiar porous α-Fe_2O_3, γ-Fe_2O_3 and Fe_3O_4 nanospheres: facile synthesis and electromagnetic properties. *Powder Technology* 269: 443–451.

61 Cheng, F. and Li, X. (2020). Effects of in situ Co or Ni doping on the photoelectrochemical performance of hematite nanorod arrays. *Applied Sciences* 10 (10): 3567.

62 Morin, F.J. (1954). Electrical properties of α-Fe_2O_3. *Physical Review* 93 (6): 1195.

63 Haynes, W. (2011). *CRC Handbook of Chemistry and Physics*, 92e. Boca Raton, FL: CRC Press/Taylor and Francis.

64 Sun, G., Dong, B., Cao, M. et al. (2011). Hierarchical dendrite-like magnetic materials of Fe_3O_4, γ-Fe_2O_3, and Fe with high performance of microwave absorption. *Chemistry of Materials* 23 (6): 1587–1593.

65 Lee, K.K., Deng, S., Fan, H.M. et al. (2012). α-Fe_2O_3 nanotubes-reduced graphene oxide composites as synergistic electrochemical capacitor materials. *Nanoscale* 4 (9): 2958–2961.

66 Hotta, M., Hayashi, M., and Nagata, K. (2011). High temperature measurement of complex permittivity and permeability of Fe_3O_4 powders in the frequency range of 0.2 to 13.5 GHz. *ISIJ International* 51 (3): 491–497.

67 Liu, X., Wing Or, S., Ming Leung, C. et al. (2013). Microwave complex permeability of Fe_3O_4 nanoflake composites with and without magnetic field-induced rotational orientation. *Journal of Applied Physics* 113 (17): 17B307.

68 Ramajo, L.A., Cristóbal, A.A., Botta, P.M. et al. (2009). Dielectric and magnetic response of Fe_3O_4/epoxy composites. *Composites Part A: Applied Science and Manufacturing* 40 (4): 388–393.

69 Radoń, A., Łukowiec, D., Kremzer, M. et al. (2018). Electrical conduction mechanism and dielectric properties of spherical shaped Fe_3O_4 nanoparticles synthesized by co-precipitation method. *Materials* 11 (5): 735.

70 Ramirez, S., Chan, K., Hernandez, R. et al. (2017). Thermal and magnetic properties of nanostructured densified ferrimagnetic composites with graphene - graphite fillers. *Materials and Design* 118: 75–80.

71 Oh, K.K., Lee, H., Lee, W.S. et al. (2009). Effect of viscosity on the magnetic permeability of Sendust-filled polymer composites. *Journal of Magnetism and Magnetic Materials* 321 (9): 1295–1299.

72 Tsutaoka, T., Ono, T., Tsurunaga, A. et al. (2011). High frequency permeability of Fe-Al-Si granular composite materials. In: *2011 IEEE International Symposium on Electromagnetic Compatibility*, 78–83. IEEE.

73 Lee, J., Nam, B., Ko, F.K. et al. (2015). Thermal conductivity prediction of magnetic composite sheet for near-field electromagnetic absorption. *AIP Journal of Applied Physics* 117 (17): 17C122.

74 Kasagi, T., Tsutaoka, T., and Hatakeyama, K. (2006). Negative permeability spectra in permalloy granular composite materials. *Applied Physics Letters* 88 (17): 172502.

75 Hengstmann, T., Grundler, D., Heyn, C. et al. (2001). Stray-field investigation on permalloy nanodisks. *AIP Journal of Applied Physics* 90 (12): 6542–6544.

76 Talaat, A., Greve, D.W., Suraj, M. et al. (2020). Electromagnetic assisted thermal processing of amorphous and nanocrystalline soft magnetic alloys: fundamentals and advances. *Journal of Alloys and Compounds* 854: 156480.

77 Babayan, V., Kazantseva, N.E., Sapurina, I. et al. (2013). Increasing the high-frequency magnetic permeability of MnZn ferrite in polyaniline composites by incorporating silver. *Journal of Magnetism and Magnetic Materials* 333: 30–38.

78 Farheen, A. and Singh, R. (2019). FMR and magnetic studies of RF-sputtered Mn-Zn ferrite thin films. *Journal of Superconductivity and Novel Magnetism* 32 (8): 2679–2686.

79 Brühl, M., Zhang, D., Talaka, T. et al. (2010). Microstructure and complex magnetic permeability of thermally sprayed NiZn ferrite coatings for electromagnetic wave absorbers. *Surface Engineering* 26 (6): 484–490.

80 Tsutaoka, T., Ueshima, M., Tokunaga, T. et al. (1995). Frequency dispersion and temperature variation of complex permeability of Ni-Zn ferrite composite materials. *AIP Journal of Applied Physics* 78 (6): 3983–3991.

81 Ali, N.N., Atassi, Y., Salloum, A. et al. (2018). Comparative study of microwave absorption characteristics of (polyaniline/NiZn ferrite) nanocomposites with different ferrite percentages. *Materials Chemistry and Physics* 211: 79–87.

82 Javed, H., Rehman, A., Mussadiq, S. et al. (2019). Reduced graphene oxide-spinel ferrite nano-hybrids as magnetically separable and recyclable visible light driven photocatalyst. *Synthetic Metals* 254: 1–9.

83 Deokar, G., Avila, J., Razado-Colambo, I. et al. (2015). Towards high quality CVD graphene growth and transfer. *Carbon* 89: 82–92.

84 Vlassiouk, I., Smirnov, S., Ivanov, I. et al. (2011). Electrical and thermal conductivity of low temperature CVD graphene: the effect of disorder. *Nanotechnology* 22 (27): 275716.

85 Sun, H., Ge, G., Zhu, J. et al. (2015). High electrical conductivity of graphene-based transparent conductive films with silver nanocomposites. *RSC Advances* 5 (130): 108044–108049.

86 Cai, W., Moore, A.L., Zhu, Y. et al. (2010). Thermal transport in suspended and supported monolayer graphene grown by chemical vapor deposition. *Nano Letters* 10 (5): 1645–1651.

87 Wang, L., Yin, M., Zhong, B. et al. (2019). Quantum transport properties of monolayer graphene with antidot lattice. *Journal of Applied Physics* 126 (8): 084305.

88 Petrone, N., Dean, C.R., Meric, I. et al. (2012). Chemical vapor deposition-derived graphene with electrical performance of exfoliated graphene. *Nano Letters* 12 (6): 2751–2756.

89 Kang, M.H., Qiu, G., Chen, B. et al. (2017). Transport in polymer-supported chemically-doped CVD graphene. *Journal of Materials Chemistry C* 5 (38): 9886–9897.

90 Venugopal, G., Krishnamoorthy, K., Mohan, R. et al. (2012). An investigation of the electrical transport properties of graphene-oxide thin films. *Materials Chemistry and Physics* 132 (1): 29–33.

91 Eda, G. and Chhowalla, M. (2010). Chemically derived graphene oxide: towards large-area thin-film electronics and optoelectronics. *Advanced Materials* 22 (22): 2392–2415.

92 Mahanta, N.K. and Abramson, A.R. (2012). Thermal conductivity of graphene and graphene oxide nanoplatelets. *13th InterSociety Conference on Thermal and Thermomechanical Phenomena in Electronic Systems.*

93 Jin, M., Jeong, H.-K., Yu, W.J. et al. (2009). Graphene oxide thin film field effect transistors without reduction. *Journal of Physics D: Applied Physics* 42 (13): 135109.

94 Chen, H., Müller, M.B., Gilmore, K.J. et al. (2008). Mechanically strong, electrically conductive, and biocompatible graphene paper. *Advanced Materials* 20 (18): 3557–3561.

95 Yang, H., Cao, Y., He, J. et al. (2017). Highly conductive free-standing reduced graphene oxide thin films for fast photoelectric devices. *Carbon* 115: 561–570.

96 Chang, H., Sun, Z., Yuan, Q. et al. (2010). Thin film field-effect phototransistors from bandgap-tunable, solution-processed, few-layer reduced graphene oxide films. *Advanced Materials* 22 (43): 4872–4876.

97 Kumar, P., Shahzad, F., Yu, S. et al. (2015). Large-area reduced graphene oxide thin film with excellent thermal conductivity and electromagnetic interference shielding effectiveness. *Carbon* 94: 494–500.

98 Gupta, S., Joshi, P., and Narayan, J. (2020). Electron mobility modulation in graphene oxide by controlling carbon melt lifetime. *Carbon* 170: 327–337.

99 Shahzad, F., Kumar, P., Kim, Y.-H. et al. (2016). Biomass-derived thermally annealed interconnected sulfur-doped graphene as a shield against electromagnetic interference. *ACS Applied Materials & Interfaces* 8 (14): 9361–9369.

100 Kajen, R., Chandrasekhar, N., Pey, K.L. et al. (2013). Charge transport in lightly reduced graphene oxide: a transport energy perspective. *Journal of Applied Physics* 113 (6): 063710.

101 Liang, C., Wang, Y., and Li, T. (2015). Synthesis of sulfur-doped p-type graphene by annealing with hydrogen sulfide. *Carbon* 82: 506–512.

102 Zhou, J., Wang, Z., Chen, Y. et al. (2017). Growth and properties of large-area sulfur-doped graphene films. *Journal of Materials Chemistry C* 5 (31): 7944–7949.

103 Denis, P.A. (2013). Concentration dependence of the band gaps of phosphorus and sulfur doped graphene. *Computational Materials Science* 67: 203–206.

104 Wang, Z., Li, P., Chen, Y. et al. (2014). Pure thiophene–sulfur doped reduced graphene oxide: synthesis, structure, and electrical properties. *Nanoscale* 6 (13): 7281–7287.

105 Usachov, D., Vilkov, O., Gruneis, A. et al. (2011). Nitrogen-doped graphene: efficient growth, structure, and electronic properties. *Nano Letters* 11 (12): 5401–5407.

106 Chang, C.-K., Kataria, S., Kuo, C.-C. et al. (2013). Band gap engineering of chemical vapor deposited graphene by in situ BN doping. *ACS Nano* 7 (2): 1333–1341.

107 Wu, T., Shen, H., Sun, L. et al. (2012). Nitrogen and boron doped monolayer graphene by chemical vapor deposition using polystyrene, urea and boric acid. *New Journal of Chemistry* 36 (6): 1385–1391.

108 Mann, S., Mudahar, I., Sharma, H. et al. (2020). Lattice thermal conductivity of pure and doped (B, N) graphene. *Materials Research Express* 7 (9): 095003.

109 Wang, H., Zhou, Y., Wu, D. et al. (2013). Synthesis of boron-doped graphene monolayers using the sole solid feedstock by chemical vapor deposition. *Small* 9 (8): 1316–1320.

110 Wang, Z., Li, P., Chen, Y. et al. (2014). Synthesis of nitrogen-doped graphene by chemical vapour deposition using melamine as the sole solid source of carbon and nitrogen. *Journal of Materials Chemistry C* 2 (35): 7396–7401.

111 Li, C., Hu, Y., Yu, M. et al. (2014). Nitrogen doped graphene paper as a highly conductive, and light-weight substrate for flexible supercapacitors. *RSC Advances* 4 (94): 51878–51883.

112 Arellano, L.M., Yue, S., Atienzar, P. et al. (2019). Modulating charge carrier density and mobility in doped graphene by covalent functionalization. *Chemical Communications* 55 (67): 9999–10002.

113 Kim, T.Y., Park, C.-H., and Marzari, N. (2016). The electronic thermal conductivity of graphene. *Nano Letters* 16 (4): 2439–2443.

114 Rajan, A.C., Mishra, A., Satsangi, S. et al. (2018). Machine-learning-assisted accurate band gap predictions of functionalized MXene. *Chemistry of Materials* 30 (12): 4031–4038.

115 Han, M., Shuck, C.E., Rakhmanov, R. et al. (2020). Beyond Ti_3C_2Tx: MXenes for electromagnetic interference shielding. *ACS Nano* 14 (4): 5008–5016.

116 Zhang, Y., Xia, W., Wu, Y. et al. (2019). Prediction of MXene based 2D tunable band gap semiconductors: GW quasiparticle calculations. *Nanoscale* 11 (9): 3993–4000.

117 Zha, X.-H., Huang, Q., He, J. et al. (2016). The thermal and electrical properties of the promising semiconductor MXene Hf_2CO_2. *Scientific Reports* 6 (1): 27971.

118 Jiang, X., Kuklin, A.V., Baev, A. et al. (2020). Two-dimensional MXenes: from morphological to optical, electric, and magnetic properties and applications. *Physics Reports* 848: 1–58.

119 Khazaei, M., Arai, M., Sasaki, T. et al. (2013). Novel electronic and magnetic properties of two-dimensional transition metal carbides and nitrides. *Advanced Functional Materials* 23 (17): 2185–2192.

120 Shao, Y., Zhang, F., Shi, X. et al. (2017). N-Functionalized MXenes: ultrahigh carrier mobility and multifunctional properties. *Physical Chemistry Chemical Physics* 19 (42): 28710–28717.

121 Huang, Y., Zhou, J., Wang, G. et al. (2019). Abnormally strong electron–phonon scattering induced unprecedented reduction in lattice thermal conductivity of two-dimensional Nb_2C. *Journal of the American Chemical Society* 141 (21): 8503–8508.

122 Zha, X.-H., Yin, J., Zhou, Y. et al. (2016). Intrinsic structural, electrical, thermal, and mechanical properties of the promising conductor Mo_2C MXene. *The Journal of Physical Chemistry C* 120 (28): 15082–15088.

123 Zha, X.-H., Zhou, J., Zhou, Y. et al. (2016). Promising electron mobility and high thermal conductivity in Sc_2CT_2 (T = F, OH) MXenes. *Nanoscale* 8 (11): 6110–6117.

124 Kobayashi, K. and Yamauchi, J. (1995). Electronic structure and scanning-tunneling-microscopy image of molybdenum dichalcogenide surfaces. *Physical Reviews B* 51 (23): 17085.

125 Yun, W.S., Han, S., Hong, S.C. et al. (2012). Thickness and strain effects on electronic structures of transition metal dichalcogenides: 2H-MX_2 semiconductors (M = Mo, W; X = S, Se, Te). *Physical Reviews B* 85 (3): 033305.

126 Worsley, M.A., Shin, S.J., Merrill, M.D. et al. (2015). Ultralow density, monolithic WS_2, MoS_2, and MoS_2/graphene aerogels. *ACS Nano* 9 (5): 4698–4705.

127 Pearce, C.I., Pattrick, R.A., and Vaughan, D.J.J.R.i.M. (2006). et al., Electrical and magnetic properties of sulfides. *Reviews in Mineralogy and Geochemistry* 61 (1): 127–180.

128 Fuhrer, M.S. and Hone, J. (2013). Measurement of mobility in dual-gated MoS_2 transistors. *Nature Nanotechnology* 8 (3): 146–147.

129 Kim, S., Hwang, W.-S., Lee, J.H. et al. (2012). High-mobility and low-power thin-film transistors based on multilayer MoS_2 crystals. *Nature Communications* 3 (1): 1–7.

130 Joo, M.-K., Moon, B.H., Ji, H. et al. (2016). Electron excess doping and effective Schottky barrier reduction on the MoS_2/h-BN heterostructure. *Nano Letters* 16 (10): 6383–6389.

131 Balach, J., Linnemann, J., Jaumann, T., and Giebeler, L.J.J.o.M.C.A. (2018). et al., Metal-based nanostructured materials for advanced lithium–sulfur batteries. *Journal of Materials Chemistry A* 6 (46): 23127–23168.

132 Velický, M. and Toth, P.S. (2017). From two-dimensional materials to their heterostructures: an electrochemist's perspective. *Applied Materials Today* 8: 68–103.

133 Eagleson, M. (1994). *Concise Encyclopedia Chemistry*. Walter de Gruyter.

134 Baglio, J.A., Calabrese, G.S., Kamieniecki, E. et al. (1982). Characterization of n-type semiconducting tungsten disulfide photoanodes in aqueous and non-aqueous electrolyte solutions: photo-oxidation of halides with high efficiency. *Journal of Electrochemical Society* 129 (7): 1461.

135 Gutiérrez, H.R., Perea-López, N., Elías, A.L. et al. (2013). Extraordinary room-temperature photoluminescence in triangular WS_2 monolayers. *Nano Letters* 13 (8): 3447–3454.

136 Zhang, C., Ning, Z., Liu, Y. et al. (2012). Electrical transport properties of individual WS_2 nanotubes and their dependence on water and oxygen absorption. *AIP Applied Physics Letters* 101 (11): 113112.

137 Peimyoo, N., Shang, J., Yang, W. et al. (2015). Thermal conductivity determination of suspended mono- and bilayer WS_2 by Raman spectroscopy. *Nano Research* 8 (4): 1210–1221.

138 Li, L., Deng, X., Wang, Z. et al. (2017). Superconducting order from disorder in 2H-$TaSe_{2-x}S_x$. *npj Quantum Materials* 2 (1): 1–7.

139 Sunilkumar, A., Manjunatha, S., Chethan, B. et al. (2019). Polypyrrole–tantalum disulfide composite: an efficient material for fabrication of room temperature operable humidity sensor. *Sensors and Actuators A: Physical* 298: 111593.

140 Hollander, M.J., Liu, Y., Lu, W.-J. et al. (2015). Electrically driven reversible insulator–metal phase transition in 1T-TaS_2. *Nano Letters* 15 (3): 1861–1866.

141 Çakıroğlu, O., Mehmood, N., Çiçek, M.M. et al. (2020). Thermal conductivity measurements in nanosheets via bolometric effect. *2D Materials* 7 (3): 035003.

142 Sun, K., Su, D., Zhang, Q. et al. (2015). Interaction of CuS and sulfur in Li-S battery system. *Journal of the Electrochemical Society* 162 (14): A2834.

143 Chloob, M.K. and Hussain, S.A. (2020). Study the structural and optical properties pure copper sulfide (CuS) films prepared by pulsed laser deposition (PLD). *Journal of Physics: Conference Series* 1591: 012014.

144 Kar, P., Farsinezhad, S., Zhang, X. et al. (2014). Anodic Cu_2S and CuS nanorod and nanowall arrays: preparation, properties and application in CO_2 photoreduction. *Nanoscale* 6 (23): 14305–14318.

145 Maneesha, P., Paulson, A., Muhammed Sabeer, N.A. et al. (2018). Thermo electric measurement of nanocrystalline cobalt doped copper sulfide for energy generation. *Materials Letters* 225: 57–61.

146 Sahoo, S.K. and Wei, K.H. (2019). A perspective on recent advances in 2D stanene nanosheets. *Advanced Materials Interfaces* 6 (18): 1900752.

147 Zhang, J., Tu, R., Goto, T.J.J.o.a. et al. (2010). Preparation of Ni-precipitated hBN powder by rotary chemical vapor deposition and its consolidation by spark plasma sintering. *Journal of Alloys and Compounds* 502 (2): 371–375.

148 Li, L., Yu, Y., Ye, G.J. et al. (2014). Black phosphorus field-effect transistors. *Nature Nanotechnology* 9 (5): 372.

149 Jain, A. and McGaughey, A.J.H. (2015). Strongly anisotropic in-plane thermal transport in single-layer black phosphorene. *Scientific Reports* 5: 8501.

150 Qiao, J., Kong, X., Hu, Z.-X. et al. (2014). et al., High-mobility transport anisotropy and linear dichroism in few-layer black phosphorus. *Nature Communications* 5 (1): 1–7.

151 Zhang, G., Huang, S., Chaves, A. et al. (2017). et al., Infrared fingerprints of few-layer black phosphorus. *Nature Communications* 8 (1): 1–9.

152 Tran, V., Soklaski, R., Liang, Y. et al. (2014). Layer-controlled band gap and anisotropic excitons in few-layer black phosphorus. *Physical Review B* 89 (23): 235319.

153 Saito, Y. and Iwasa, Y. (2015). Ambipolar insulator-to-metal transition in black phosphorus by ionic-liquid gating. *ACS Nano* 9 (3): 3192–3198.

154 Liu, C., Li, H., Xu, H. et al. (2019). Slow light effect based on tunable plasmon-induced transparency of monolayer black phosphorus. *Journal of Physics D: Applied Physics* 52 (40): 405203.

155 Low, T., Engel, M., Steiner, M. et al. (2014). Origin of photoresponse in black phosphorus phototransistors. *Physical Review B* 90 (8): 081408.

156 Liu, H., Du, Y., Deng, Y. et al. (2015). Semiconducting black phosphorus: synthesis, transport properties and electronic applications. *Chemical Society Reviews* 44 (9): 2732–2743.

2

EMI Shielding Mechanism and Conversion Techniques

2.1 Introduction

To design a shielding material, one has to understand the basic electromagnetic interference (EMI) shielding mechanisms and how an electromagnetic wave (EMW) interacts with the shielding material. To assess the EMI shielding efficiency, the conductivity, permittivity, and permeability of the shield should be considered. Scattering parameter conversion techniques, such as the Nicholson–Ross–Weir (NRW) method, the National Institute of Standards and Technology (NIST) iterative method, the new non-iterative method, and the short-circuit line (SCL) method, are necessary to extract these attributes of the shielding material from the scattering parameters measured using a vector network analyzer (VNA). Every conversion technique has advantages and shortcomings, and the selection of an appropriate conversion technique is important to achieve accuracy. This chapter provides a theoretical overview of the basic EMI shielding mechanisms, the scattering parameter conversion methods, and the effect of various parameters on the EMI shielding performance.

2.2 EMI Shielding Mechanisms

Shielding technology aims to minimize the transmission of EMWs by controlling reflection, absorption, and multiple reflections within a shield (Figure 2.1). Reflection is the primary shielding mechanism. For EMWs incident on the surface of the shield, some of the EMWs are reflected from the surface because of the impedance mismatch between air and the shield. Absorption loss is the second shielding mechanism. The EMWs that are not reflected cross the front surface of the shield and start to propagate inside the shield. During propagation, the intensity of the EMWs is attenuated exponentially as the shield thickness increases. Attenuation by absorption is achieved through Ohmic loss, dielectric loss, and magnetic loss. Some of the EMWs that reach the back surface of the shield are also reflected, and a small portion of the EMWs is finally transmitted through the shield. The shielding performance is also influenced by the multiple reflections, which arise because of repeated reflections within the shield. This phenomenon occurs within very thin shields or shields

Two-Dimensional Materials for Electromagnetic Shielding, First Edition.
Chong Min Koo, Pradeep Sambyal, Aamir Iqbal, Faisal Shahzad, and Junpyo Hong.

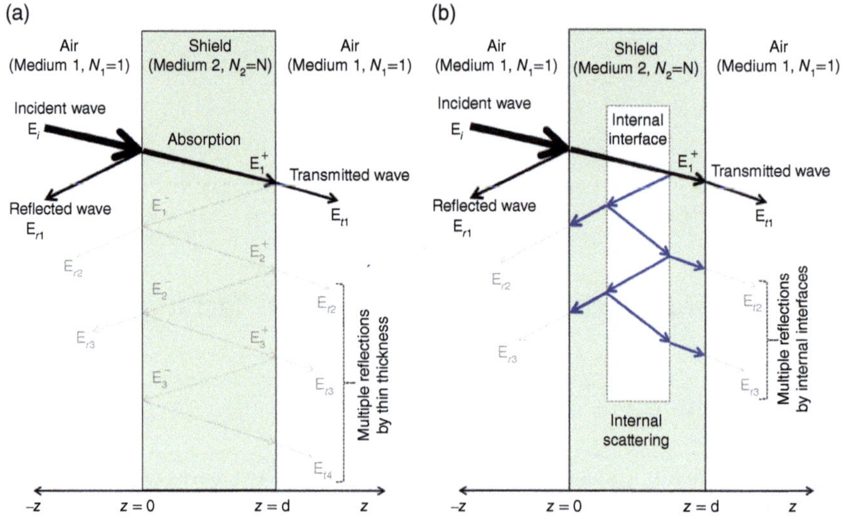

Figure 2.1 Schematic representations of the propagation of EMWs with multiple reflections caused by (a) a thickness smaller than the skin depth and (b) internal interfaces (pores, cavities, or a secondary phase). Source: Chong Min Koo.

with foam, porous, or honeycomb structures, where the EMWs are scattered at the internal interfaces within the shield [1]. Both of these multiple reflections vary in their mechanisms (see Figure 2.1).

2.2.1 Shielding Effectiveness (SE)

When an EMW incidents on the shield, it can be reflected, absorbed, or transmitted. The electric and magnetic fields of EMWs are perpendicular to each other as well as to the propagation direction in the medium. The electric and magnetic fields of EMWs can be expressed in the phasor form as follows [2, 3]:

$$\vec{E} = Ee^{-\gamma z}\,\hat{a}_x = Ee^{-\alpha z}\,e^{-j\beta z}\hat{a}_x = Ee^{-\alpha z}\,(\cos\beta z - j\sin\beta z)\hat{a}_x \tag{2.1}$$

$$\vec{H} = He^{-\gamma z}\,\hat{a}_y = He^{-\alpha z}e^{-j\beta z}\hat{a}_y = He^{-\alpha z}\,(\cos\beta z - j\sin\beta z)\hat{a}_y \tag{2.2}$$

where γ is the propagation constant of the medium, α and β are the phase shift and attenuation constants, respectively, and E and H denote the electric and magnetic field amplitudes, respectively. In a single medium, EMWs do not undergo reflection, whereas when propagating in two different media, reflection may occur. Thus, when EMW incident from medium 1 to medium 2 direction, the reflection coefficient (r_{12}) and transmission coefficient (t_{12}) at the interface between medium 1 and medium 2 can be obtained from the impedance of the medium as follows [2, 3]:

$$r_{12} = \frac{E_r}{E_i} = \frac{\eta_2 - \eta_1}{\eta_2 + \eta_1} = \frac{n_1 - n_2}{n_1 + n_2} \tag{2.3}$$

$$t_{12} = \frac{E_t}{E_i} = \frac{2\eta_2}{\eta_2 + \eta_1} = \frac{2n_1}{n_1 + n_2} \tag{2.4}$$

where the amplitudes of the incident and reflected electric fields are denoted by E_i and E_r, respectively, and the impedances of media 1 and 2 are represented by η_1 and η_2, respectively, and the refractive index of media 1 and 2 are represented by n_1 and n_2, respectively. This relationship indicates that reflection occurs at interfaces between media with different impedances. The impedance of a medium can be expressed as the ratio of the electric field to the magnetic field and thus can be defined as a function of four parameters (electrical conductivity [σ], magnetic permeability [μ], permittivity [ε], and angular frequency [ω]) as follows:

$$\eta = \frac{|E|}{|H|} = \sqrt{\frac{j\omega\mu}{\sigma + j\omega\varepsilon}} \tag{2.5}$$

According to plane wave theory, absorption occurs during the propagation of EMWs in a medium. The amplitude of an EMW decreases exponentially as it passes through a shielding material with an attenuation constant of α, as shown in Figure 2.1. This exponential decrease in the strength or amplitude (E) of the EMW occurs because $E = E_0 e^{-\alpha z}$ and $H = H_0 e^{-\alpha z}$. The attenuation constant (α) can be defined as a function of the parameters of the shield (ω, μ, σ, and ε) according to Eq. (2.6) [2, 3]:

$$\alpha = \omega\sqrt{\frac{\mu\varepsilon}{2}\left[\sqrt{1 + \left(\frac{\sigma}{\omega\varepsilon}\right)^2} - 1\right]} \tag{2.6}$$

Consequently, EMWs should undergo reflection at both the front and back surfaces of the shield as well as absorption in the shield before transmission (Figure 2.1a). The ratio of the total transmitted field relative to the incident field can be defined as follows [2, 3]:

$$\frac{E_t}{E_i} = t_{12}e^{-\alpha d}t_{21} = \frac{2\eta}{\eta + \eta_0}e^{-\alpha d}\frac{2\eta_0}{\eta + \eta_0} = \frac{4\eta\eta_0}{(\eta + \eta_0)^2}e^{-\alpha d} \tag{2.7}$$

Given the transmission to incident ratio of the intensity of the EMWs, each transmission and reflection part is considered as:

$$\frac{Et_1}{E_i} = t_{12}\,e^{-\alpha d}\,t_{21} \tag{2.8}$$

$$\frac{Et_2}{E_i} = t_{12}\,e^{-\alpha d}\,r_{21}\,e^{-\alpha d}\,r_{21}\,e^{-\alpha d}\,t_{21} \tag{2.9}$$

$$\frac{Et_3}{E_i} = t_{12}\,e^{-\alpha d}\,r_{21}\,e^{-\alpha d}\,r_{21}\,e^{-\alpha d}\,r_{21}\,e^{-\alpha d}\,r_{21}\,e^{-\alpha d}\,t_{21} \tag{2.10}$$

$$\frac{E_t}{E_i} = \frac{Et_1 + Et_2 + Et_3 + \dots}{E_i} = t_{12}e^{-\alpha d}\,t_{21}[1 + r_{21}^2\,e^{-2\alpha d} + (r_{21}^2\,e^{-2\alpha d})^2 + \dots] \tag{2.11}$$

When $1 < x < 1$, then

$$1 + x + x^2 + x^3 + \cdots = \frac{1}{1 - x} \tag{2.12}$$

Therefore,

$$\frac{E_t}{E_i} = t_{12}e^{-\alpha d}t_{21}.\frac{1}{1-r_{21}^2\,e^{-2\alpha d}} = \frac{4n_1n_2}{(n_2+n_1)^2}e^{-\alpha d}.\frac{1}{1-\left(\frac{n_2-n_1}{n_2+n_1}\right)^2 e^{-2\alpha d}}$$

$$= \frac{4\eta_2\eta_1}{(\eta_2+\eta_1)^2}e^{-\alpha d}.\frac{1}{1-\left(\frac{\eta_1-\eta_2}{\eta_1+\eta_2}\right)^2 e^{-2\alpha d}} \tag{2.13}$$

Shielding efficiency is quantified by shielding effectiveness (SE) that is defined as the logarithmic ratio of the power of incident and transmitted EMWs. The power of EMW represents the square of the electric field amplitude.

$$\mathrm{SE}_T = 20\log\left(\frac{E_i}{E_t}\right)^2 = 10\log\frac{1}{|t|^2} = 20\log\frac{1}{|\,t\,|} \tag{2.14}$$

$$\mathrm{SE}_T = 20\log\left[\frac{(\eta_2+\eta_1)^2}{4\eta_2\eta_1}\right] + 20\log e^{\alpha d} + 20\log\left[1-\left(\frac{\eta_1-\eta_2}{\eta_1+\eta_2}\right)^2 e^{-2\alpha d}\right] \tag{2.15}$$

or,

$$\mathrm{SE}_T = 20\log\left[\frac{(n_2+n_1)^2}{4n_2n_1}\right] + 20\log e^{\alpha d} + 20\log\left[1-\left(\frac{n_2-n_1}{n_2+n_1}\right)^2 e^{-2\alpha d}\right] \tag{2.16}$$

Total EMI SE SE_T can be expressed by the combination of shielding by reflection SE_R, shielding by absorption SE_A, and multiple reflections SE_M.

$$\mathrm{SE}_T = \mathrm{SE}_R + \mathrm{SE}_A + \mathrm{SE}_M \tag{2.17}$$

SE_R arises due to the reflection from the surface of the shield, while SE_A signifies absorption of EMW within the shield.

When a material is electrically conductive, SE_R, SE_A, and SE_M can be simplified to:

$$\mathrm{SE}_R = 20\log\left[\frac{(\eta_2+\eta_1)^2}{4\eta_2\eta_1}\right] = 20\log\left[\frac{(n_2+n_1)^2}{4n_2n_1}\right] = 20\log\left[\frac{\eta_0}{4\eta_2}\right] \tag{2.18}$$

$$\mathrm{SE}_R = 39.5 + 10\log\left[\frac{\sigma}{2\pi f\mu}\right] = 50 + 10\log\left[\frac{\sigma}{f}\right] \tag{2.19}$$

$$\mathrm{SE}_A = 20\log e^{\alpha d} = 20\,\alpha d\,\log_{10}e = 8.7\,\alpha d = 8.7\,\frac{d}{\delta} = 8.7d\sqrt{\pi f\mu\sigma} = 1.7\,d\,\sqrt{\sigma f} \tag{2.20}$$

$$\mathrm{SE}_M(\mathrm{dB}) = 20\log_{10}(1-e^{-2\alpha d}) = 20\log_{10}\left(1-e^{-\frac{2d}{\delta}}\right) \tag{2.21}$$

Above skin depth, the SE_M becomes negligible. The SE SE_T is expressed with the combination of SE_R and SE_A.

$$\mathrm{SE}_T = 50 + 10\log\left[\frac{\sigma}{f}\right] + 1.7\,d\,\sqrt{\sigma f} \tag{2.22}$$

The above equation is known as Simon's formula.

Multiple reflections should be considered at very thin thickness. When the incident EMW enter a thin shield, the reflected radiations from the back surface are re-reflected to the front surface and then reflected to the back surface (Figure 2.1a). This leads to additional propagation of EMW at every single reflection moment, and this phenomenon remains to continue until the energy of the reflected EMW is completely attenuated. This mechanism is called multiple reflections, and it happens at very thin thickness, lower than the skin depth of the material [3, 4]. The SE_M is greatly reliant on the thickness of the shield and becomes insignificant at a thickness close to or greater than the skin depth or when the SE_A value reaches higher than 10 dB. Conversely, if the thickness is less than the skin depth, multiple reflections must be considered when determining the SE. It is noteworthy that the multiple reflections due to a thin thickness have a negative impact on the SE_T value.

Strong multiple reflections can also occur from internal interfaces present within a shield. Internal inclusions such as dielectric domains or pores enhance internal scattering (also called multiple reflections) within the shield, consequently attenuating the power of EMWs (as shown in Figure 2.1b). In contrast to the previous one, this form of multiple reflections has a positive impact on the SE_A and SE_T values. In particular, this mechanism enhances the absorption of EMWs inside the shield and thus the total EMI SE.

2.2.2 SE/*t*, SSE, and SSE/*t* for Lightweight Shielding Materials with Minimal Thicknesses

Lightweight shielding materials with minimal thicknesses are critical for highly integrated electronic devices in mobile and aerospace applications. The parameters SE/t, SSE, and SSE/t have been defined to assess the performance of shielding materials in these applications. SE/t is the EMI SE normalized by the thickness (t) of the shield. The EMI SE value normalized by the density (ρ) gives the specific shielding effectiveness (SSE). Finally, the EMI SE value normalized by the density (ρ) and the thickness (t) gives the absolute shielding effectiveness (SSE/t), where a higher value indicates better performance for a lightweight material with a minimal thickness in next-generation smart and portable electronics. Mathematically, SSE/t is expressed as follows:

$$\mathrm{SSE}/t = \mathrm{EMI\,SE}/\rho/t = \mathrm{dB\,cm^2\,g^{-1}} \tag{2.23}$$

2.2.3 Impact of Different Parameters on Electromagnetic Shielding Effectiveness

The shielding efficiency of a material depends on various parameters, and minute variations in one factor can affect the other parameters and lead to variations in the final EMI SE value. This section provides a brief discussion on the various factors that directly or indirectly affect the calculations of EMI SE.

2.2.3.1 Distance of Shield from the Source

The electromagnetic radiative region can be divided into three parts based on the distance r between the radiative source and the shield. The near- and far-field regions are located at $r < \lambda/2\pi$ and $r > \lambda/2\pi$, respectively, and the space between

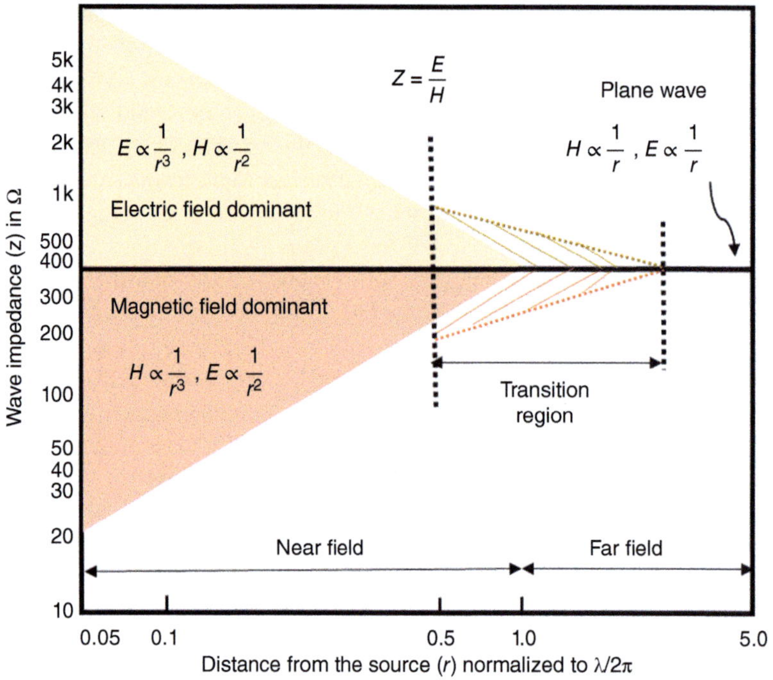

Figure 2.2 Dependence of wave impedance on the distance from the source (normalized to $\lambda/2\pi$). Source: Chong Min Koo.

these regions is the transition region ($r \approx \lambda/2\pi$), as shown in Figure 2.2. In the near field, electric field dominant source and the magnetic dominant source produce the different wave impedance behavior. In case of magnetic source, the electric field diminishes at a rate of $(1/r)^2$, while the magnetic field diminishes at a rate of $(1/r)^3$. In case of electric source, the electric field reduces at a rate of $(1/r)^3$, while the magnetic field reduces at a rate of $(1/r)^2$. In the far field, both the electric and magnetic fields diminish at a same rate of $(1/r)$ [5]. Therefore, the impedance of the EMWs is constant in the far field, where they act as plane waves. This book deals with the measurement and analysis of the EMI SE of materials in the far field.

2.2.3.2 Frequency of the Incident Electromagnetic Field

The shielding by reflection (SE_R) and by absorption (SE_A) depends on the frequency, as shown in Eqs. (2.19) and (2.20). SE_R is inversely proportional to the frequency, i.e. it decreases with increasing frequency. This reduction in SE_R is due to the increase in shield impedance with increasing frequency. In contrast, SE_A is directly proportional to the frequency, owing to the decrease in skin depth with increasing frequency.

2.2.3.3 Electrical Conductivity or Sheet Resistance

Electrical conductivity is an important parameter, which directly defines the EMI shielding ability of any material, i.e. the higher the conductivity, higher the EMI SE, and vice versa. Higher electrical conductivity like in case of metals results in a reflection-dominant EMI shielding mechanism, which is set to make absorption dominant by controlling the electrical conductivity. As given in details in Section 2.1, SE_R and SE_A are directly related to the electrical conductivity, whereas SE_M is indirectly affected by this factor, which alters the skin depth of the shielding material (see Eqs. (2.19)–(2.21)). In real application, sheet resistance (R_s) is a more convenient parameter than conductivity. At ultrathin thickness, sheet resistance (R_s) can be used to calculate transmission and reflection coefficients as [4, 6]:

$$T = |t|^2 = \left|\frac{1}{(1+Z_0)/(2R_s\cos\theta_0)}\right|^2 \tag{2.24}$$

$$R = |r|^2 = \left|\frac{-Z_0/(2R_s\cos\theta_0)}{(1+Z_0)/(2R_s\cos\theta_0)}\right|^2 \tag{2.25}$$

2.2.3.4 Thickness of Shield

SE_R is independent of the shield thickness, whereas SE_A and SE_M vary directly with increasing shield thickness (Eqs. 2.19–2.21). The thickness of a material increases the propagation path length for the EMW to travel before transmission. Therefore, an increased thickness enhances the shielding by absorption (SE_A). In addition, multiple reflections are also highly affected by the thickness of the material. As explained in Section 2.1 and Figure 2.1a, electrically thin films (where the thickness of the film is smaller than the skin depth) exhibit strong multiple reflections, but this phenomenon is not effective at greater thicknesses (larger than the skin depth).

By combining the effect of electrical conductivity and thickness, the obtained function is graphically represented as (Figure 2.3).

2.2.3.5 Dielectric Losses

The electromagnetic absorption characteristics of a shield depend on the dielectric properties represented by the complex permittivity (ε', ε''). The real part (ε') of the complex permittivity is related to the polarization within the material, resulting in capacitive energy storage, whereas the imaginary part (ε'') is associated with the dissipation of energy. The dielectric loss mechanism is governed by conduction loss, dielectric relaxation loss, resonance loss, and other losses.

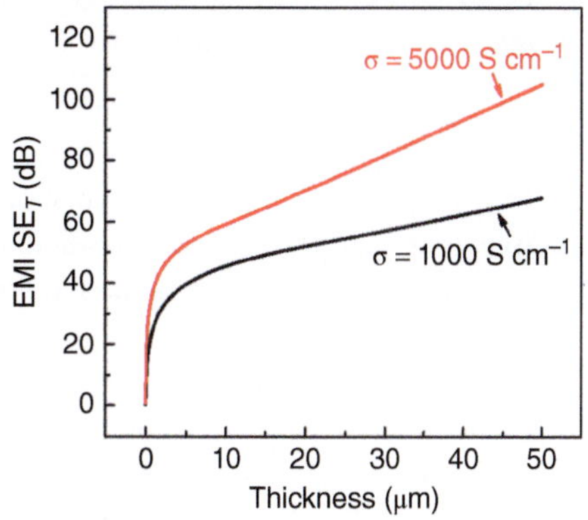

Figure 2.3 EMI SE_T as a function of electrical conductivity and thickness of a shield. Source: Chong Min Koo.

Ohmic (or Conduction) Loss Under the applied electric field of EMWs, a lossy shielding material generates a conductance current that dissipates the energy of the incident field in the form of heat. Therefore, the conductance loss of a shielding material is related to its electrical conductivity and is calculated as the conductance loss tangent [7]:

$$tg\delta_c = 1.8 \times 10^{10} \frac{\sigma}{f_{\varepsilon_r}} \tag{2.26}$$

Dielectric Relaxation Loss The dipoles in a material are polarized under the applied electric field of EMWs. The net-induced polarization can take place in different forms, including electronic polarization, ionic polarization, thermal ion polarization, dipolar (or orientation) polarization, and space charge or interfacial polarization. The type of polarization strongly depends on the frequency of the electric field and the mass of the polarized species. Ionic and electronic polarizations only occur at higher frequencies ($\sim 10^{14}$ to $10^{15}\ s^{-1}$). However, dipolar, thermal ion displacement, and interfacial polarizations can occur at frequencies in the gigahertz range and produce higher energy losses. The time lapse between the alternating electric field and the induced polarization generates dielectric relaxation loss, which is expressed by the dielectric relaxation loss tangent as [7]:

$$tg\delta_{\text{rel}} = \frac{\varepsilon_r''(\omega)}{\varepsilon_r'(\omega)} = \frac{(\varepsilon_{\text{rs}} - \varepsilon_{r\infty})\omega\tau}{\varepsilon_{\text{rs}} + \varepsilon_{r\infty}\omega^2\tau^2} \tag{2.27}$$

where ε_{rs}, and $\varepsilon_{\text{r}\infty}$ represent the permittivity at the frequency near zero and the infinity, respectively, and τ is the relaxation time.

Polarization/Resonance Loss Under an applied electric field, a resonance effect is induced in a shielding material by the vibrational movement of atoms, ions, and/or electrons [7]. This phenomenon usually occurs in the infrared to ultraviolet frequency region.

2.2.3.6 Magnetic Losses

The electromagnetic absorption characteristics of a shield also depend on the magnetic losses, as represented by the complex permeability (μ', μ''). The combined effect of eddy current loss, magnetic hysteresis loss, natural resonance, and residual losses can be expressed in the form of magnetic loss tangent as follows [8]:

$$\frac{2\pi tg\delta_m}{\mu} = ef + aB + c \tag{2.28}$$

where $tg\delta_m$, μ, e, a, B, and c correspond to the magnetic loss tangent, magnetic permeability, eddy current loss coefficient, magnetic hysteresis loss coefficient, magnetic flux density, and residual losses, respectively.

Eddy Current Loss Under an alternating magnetic field, a current is induced in magnetic materials that aids in dissipating the energy of the field in the form of heat. The eddy current loss coefficient (e) strongly depends on the thickness, conductivity, and permeability of the material. Eddy current loss is also affected by the grain size, domain orientation, morphology, and surface roughness of the material. However, materials with larger thicknesses and higher conductivities will exhibit higher eddy current losses, as given by

$$e = (4\pi^2 d^2 \mu_0 \sigma)/3 \tag{2.29}$$

Magnetic Hysteresis Loss Magnetic materials show magnetization behavior under an alternating magnetic field owing to the irreversible movement of the domains. The magnetic hysteresis loss coefficient is given as

$$a = \frac{8b}{3\mu^3 \mu_0} \tag{2.30}$$

Here, a is the hysteresis loss coefficient, b is the Rayleigh constant, μ_0 is the vacuum permeability, and μ is the permeability of the material. This equation shows that the magnetic hysteresis loss depends on the magnetic permeability and the Rayleigh constant.

2.3 Microwave Absorption Mechanisms

The EMI shielding mechanism generally depends on two contributions, first from the reflection that depends on the surface mobile charges and the second from the absorption inside the material. The reflection part depends heavily on the impedance mismatch between the material and the EMW traveling medium. The larger the difference in impedance between two mediums, the larger is the reflection from the surface. To achieve high reflection, it is therefore recommended to use materials with high electrical conductivity. For the case of absorption, the material need not be superconductive; instead, it shall possess a moderate conductivity just enough to create conducive channels. While designing the microwave absorbing (MWA) materials, conductivity is not the primary tuning parameter because reflection has to be minimal. These sorts of applications require that all the incoming EMWs get absorbed inside the material without being reflected from the surface. MWA materials are designed such that the impedance mismatching shall be minimal so that the incoming EMWs can penetrate the material. For this, a material needs to have strong dielectric permittivity and permeability that can absorb or dissipate the incoming EMWs in the form of heat. The MWA ability of a material is evaluated by the term "reflection loss (RL)" or "reflection coefficient (RC)" and attenuation constant α. The experimental measurement method of RL is different from that of the EMI SE of a material. RL is measured on the assumption of zero transmission. Therefore, a highly conductive metal foil with minimum transmission is placed as a backing support to test the absorbing material. In this way, all the incident EMW are reflected and/or absorbed, without transmission. At air–absorber–metal interfaces, wave is shifted into two components by a phase difference of 180 °C and cancels the effect of each other. RL is expressed as (Figure 2.4) [3]:

$$\mathrm{RL} = 20 \log \left| \frac{Z_{\mathrm{in}} - Z_0}{Z_{\mathrm{in}} + Z_0} \right| \tag{2.31}$$

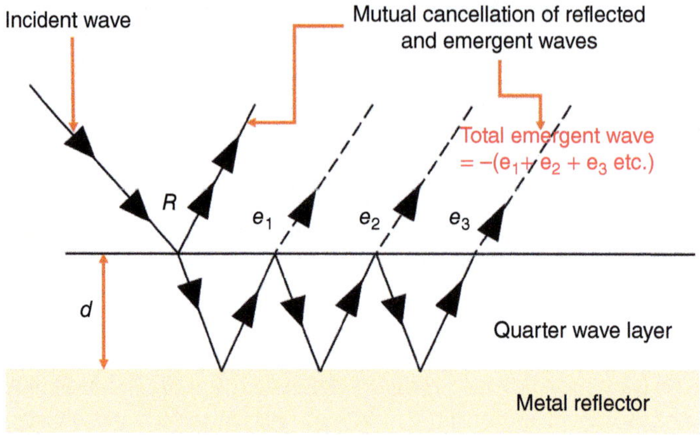

Figure 2.4 Resonant absorber showing out-of-phase condition b/w reflected and emergent wave. Source: Chong Min Koo.

In the above equation, $Z_{\text{in}}$is the characteristic input impedance of the material and $Z_0$is impedance of air. $Z_{\text{in}}$is calculated as follows [2, 3]:

$$Z_{\text{in}} = \sqrt{\frac{\mu}{\varepsilon}} \tan \text{h} \left[j \left(\frac{2\pi f d}{c} \right) \sqrt{\mu\varepsilon} \right] \tag{2.32}$$

where ε and μ are the complex permittivity and permeability of the absorbing material, respectively, while c is the speed of light.

Impedance matching at the interface between air and the lossy conducting shield ($Z_{\text{in}} \approx Z_0$) minimizes the reflection of the EMWs from the surface and is the primary requirement for minimum RL value. It also allows maximum EMWs to enter into the shield. Once the EMWs enter the shield, a material with larger attenuation constant exhibits maximum energy loss through absorption. As a negligibly low electrical conductivity is required for an absorber, the attenuation constant α can be calculated as [9, 10]:

$$\alpha = \frac{\sqrt{2}\pi f}{c} \times \sqrt{(\mu''\varepsilon'' - \mu'\varepsilon') + \sqrt{(\mu'\varepsilon'' + \mu''\varepsilon')^2 + (\mu''\varepsilon'' - \mu'\varepsilon')^2}} \tag{2.33}$$

The above equation shows that attenuation constant is directly associated with the values of complex dielectric permittivity ($\varepsilon = \varepsilon' - j\varepsilon''$) and magnetic permeability ($\mu = \mu' - j\mu''$). The real parts of complex permittivity (ε') and permeability (μ') indicate the capacitive energy storage of the electric and magnetic field, whereas the imaginary parts (ε'') and (μ'') characterize all kinds of polarization losses in the form of heat dissipation. For better understanding, tangents of dielectric loss ($\tan \delta_\varepsilon = \frac{\varepsilon''}{\varepsilon'}$) and magnetic loss ($\tan \delta_\mu = \frac{\mu''}{\mu'}$) are presented to show the loss mechanisms.

When RL curves are plotted against frequency, there are sharp peak patterns showing the dependency on the frequency. For this, a quarter-wavelength matching model needs to be considered. In fact, the peak frequency also called matching frequency (f_m) relates with the matching thickness (t_m) of the absorber and can be described as [11]:

$$t_m = \frac{n\lambda}{4} = \frac{nc}{(4f_m\sqrt{(|\varepsilon_r||\mu_r|)})} \tag{2.34}$$

where n is an integer ($n = 1, 2, 3, \ldots$) and λ is the wavelength of the EMWs. $|\varepsilon_r|$ and $|\mu_r|$ show the relative permittivity and permeability at f_m, respectively. At matching conditions of t_m and f_m, the EMWs at the air–absorber interface are out of phase to the absorber–metal interface by 180°. In this scenario, there will be extinction of the EMWs at the air–absorber interface, whereas maximum absorption will be observed within the shield at f_m. The calculated RL value of −20 dB is considered to attenuate 99% of the EMWs by absorption mechanism. The efficient absorbing materials convert the absorbed energy of EMWs into thermal energy through magnetic loss and/or dielectric loss by balancing the relative permeability and/or permittivity.

The penetrating EMWs then interact with the host structure and undergo absorption by different mechanisms such as eddy current losses, ohmic losses, polarization losses, multiple internal reflection, etc. Thus, the two mechanisms fundamentally differ from each other. The microwave absorbers tend to be less conductive but carry more magnetic constituent so that absorption of EMWs could be realized. A reasonable electrical conductivity is certainly required to provide a conduction medium. To produce a high-performance electromagnetic wave absorber (EMWA), generally the magnetic materials are used, such as sendust, ferrites, iron oxide, etc., and the conductivity part is supplemented by the addition of some metallic filler. The metallic fillers are, however, heavier and also difficult to process. Thus, the carbon-based fillers or 2D nanomaterials are being explored to not only reduce the weight of the EMWA but also the processability and cost to make market competitive. In this direction, several researchers have used different variants of carbon including graphite, graphene, and other 2D materials.

2.4 Scattering Parameter Conversion Method for Calculation of Permeability and Permittivity

2.4.1 Transmission/Reflection Method

In the transmission/reflection method, the material under test (MUT) is placed between the transmission line segment, and the permittivity and permeability of the sample are determined from the reflection and transmission of the MUT. A systematic and descriptive analysis of the working principle of the transmission/reflection method has been reported by Baker-Jarvis et al. [12–14] Herein, we briefly discuss the basic principle.

Figure 2.5 shows a characteristic measurement arrangement for the transmission/reflection method. The MUT is placed within the transmission line segment, which has an axis in the x-direction. Scattering equations are obtained by analyzing the electric field at the MUT interfaces. We assume that E_{I}, E_{II}, and E_{III} represent the electric fields of the three segments of the transmission line. For a normalized

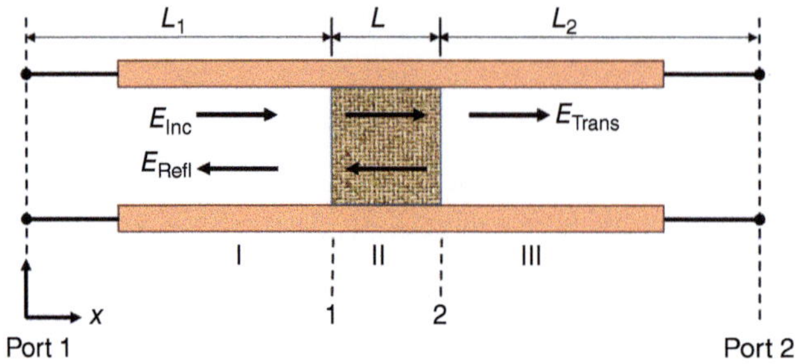

Figure 2.5 Reflection and transmission of EMWs through the transmission line. Source: Chong Min Koo.

incident wave, the spatial distribution of the electric field can be written as follows:

$$E_{\text{I}} = \exp(-\gamma_0 x) + C_1 \exp(\gamma_0 x) \tag{2.35}$$

$$E_{\text{II}} = C_2 \exp(-\gamma x) + C_3 \exp(\gamma x) \tag{2.36}$$

$$E_{\text{III}} = C_4 \exp(-\gamma_0 x) \tag{2.37}$$

where

$$\gamma = j\sqrt{\frac{\omega^2 \mu_r \varepsilon_r}{c^2} - \left(\frac{2\pi}{\lambda_c}\right)^2} \tag{2.38}$$

$$\gamma_0 = j\sqrt{\left(\frac{\omega}{c}\right)^2 - \left(\frac{2\pi}{\lambda_c}\right)^2} \tag{2.39}$$

Here, c is the speed of light in vacuum, γ_0 and γ are the propagation constants in the transmission line filled with free space and the sample, respectively, ω is the angular frequency, and λ_c is the cutoff wavelength. Constants C_i ($i = 1, 2, 3, 4$) in Eqs. (2.35)–(2.37) can be obtained from the boundary conditions on the electric and magnetic fields. The boundary condition on the electric field is the continuity of the tangential component at the interfaces. Maxwell's equations have been used to obtain the tangential component of an electric field with only a radial component.

$$E_{\text{I}}|_{x=L_1} = E_{\text{II}}|_{x=L_1} \tag{2.40}$$

$$E_{\text{II}}|_{x=L_1+L} = E_{\text{III}}|_{x=L_{1+L}} \tag{2.41}$$

where L denotes the sample length and L_1 and L_2 are the distances between the sample faces and the respective ports. The total length of the transmission line is represented by $L_{\text{air}} = L_1 + L_2 + L$. An additional assumption that no surface current is generated is also taken into account for the boundary condition on the magnetic field. If this condition holds true, the tangential component of the magnetic field is constant throughout the interface:

$$\frac{1}{\mu_0} \cdot \frac{\partial E_{\text{I}}}{\partial x}\bigg|_{x=L_1} = \frac{1}{\mu_0 \mu_r} \cdot \frac{\partial E_{\text{II}}}{\partial x}\bigg|_{x=L_1} \tag{2.42}$$

$$\frac{1}{\mu_0\mu_r}\cdot\frac{\partial E_{\text{II}}}{\partial x}\bigg|_{x=(L_1+L)} = \frac{1}{\mu_0}\cdot\frac{\partial E_{\text{III}}}{\partial x}\bigg|_{x=(L_1+L)} \tag{2.43}$$

The scattering parameters of a two-port device can be attained by solving Eqs. (2.35)–(2.37) and subjecting them to the boundary conditions (Eqs. (2.40)–(2.43)). Assuming that $S_{12} = S_{21}$, the S-parameters are defined in terms of the transmission and reflection coefficients as follows:

$$S_{11} = R_1^2.\left[\frac{\Gamma(1-T^2)}{1-\Gamma^2T^2}\right] \tag{2.44}$$

$$S_{22} = R_2^2.\left[\frac{\Gamma(1-T^2)}{1-\Gamma^2T^2}\right] \tag{2.45}$$

$$S_{21} = R_1R_2.\left[\frac{\Gamma(1-T^2)}{1-\Gamma^2T^2}\right] \tag{2.46}$$

Here, R_1 and R_2 represent the respective reference plane transformations:

$$R_1 = \exp(-\gamma_0L_1) \tag{2.47}$$

$$R_2 = \exp(-\gamma_0L_2) \tag{2.48}$$

The detailed derivation of Eqs. (2.44)–(2.46) is reported elsewhere [15, 16]. The transmission coefficient can be expressed as

$$T = \exp(-\gamma L) \tag{2.49}$$

The reflection coefficient is defined as follows:

$$\Gamma = \frac{(\gamma_0/\mu_0) - (\gamma/\mu)}{(\gamma_0/\mu_0) + (\gamma/\mu)} \tag{2.50}$$

For coaxial line, the cutoff wavelength is infinity; therefore, reflection coefficient can be written as follows:

$$\Gamma = \frac{\sqrt{\mu_r/\varepsilon_r} - 1}{\sqrt{\mu_r/\varepsilon_r} + 1} \tag{2.51}$$

Additionally, S_{21}^0 for the empty sample holder is given as

$$S_{21}^0 = R_1R_2\exp(-\gamma_0L) \tag{2.52}$$

Equations (2.44)–(2.46) for nonmagnetic materials contain ε_r', ε_r'', L, and the R_1 and R_2 reference plane transformations as unknown quantities. For this analysis, as there are four complex equations (Eqs. (2.44)–(2.46) and (2.52)) and the equation for the length of the air line, there are nine real equations for the five unknowns. Additionally, in certain applications, the sample length is known to the user. For magnetic materials, there are seven unknown quantities. As the system of equations is overdetermined, various combinations can be used to solve the equations. Thus, different methods can be employed to attain the relative permittivity (ε_r) and relative permeability (μ_r) of the MUT.

2.4.2 Nicolson–Ross–Weir (NRW) Method

The permittivity and permeability of a material can be determined from the S-parameters through direct calculations using the NRW conversion method [15, 17]. This method is primarily applicable to waveguides and coaxial line measurement systems and is generally used to carry out transformations for such systems. The reflection coefficient and transmission coefficient of the MUT can be obtained using all (S_{11}, S_{21}, S_{12}, and S_{22}) or a couple (S_{11} and S_{21}) of the S-parameters. However, this method shows deviations for low-loss materials at certain frequencies, analogous to the integer multiple of one-half wavelength in the MUT, which is attributed to phase ambiguity. Thus, it is limited to an ideal thickness of $\lambda_g/4$ and should only be utilized for short samples.

The methodology of the NRW technique is based on the following expressions:

$$S_{11} = \frac{\Gamma(1-T^2)}{(1-\Gamma^2 T^2)} \text{ and } S_{21} = \frac{T(1-\Gamma^2)}{(1-\Gamma^2 T^2)} \tag{2.53}$$

The above parameters can be attained using a VNA.

The reflection coefficient can be deduced as follows:

$$\Gamma = X \pm \sqrt{X^2 - 1} \tag{2.54}$$

where $|\Gamma_1| < 1$ is required to obtain the accurate root in terms of the S-parameter:

$$X = \frac{S_{11}^2 - S_{21}^2 + 1}{2S_{11}} \tag{2.55}$$

Similarly, the transmission coefficient can be written as follows:

$$T = \frac{S_{11} + S_{21} - \Gamma}{1 - (S_{11} + S_{21})\Gamma} \tag{2.56}$$

The permeability is given as:

$$\mu_r = \frac{1 + \Gamma_1}{\Lambda(1-\Gamma)\sqrt{\frac{1}{\lambda_0^2} - \frac{1}{\lambda_c^2}}} \tag{2.57}$$

where λ_0 is the free space wavelength, λ_c is the cutoff wavelength of the transmission line section, and $\lambda_c = \infty$ for the coaxial line.

$$\frac{1}{\Lambda^2} = \left(\frac{\varepsilon_r \mu_r}{\lambda_0^2} - \frac{1}{\lambda_c^2}\right) = -\left(\frac{1}{2\pi L} \ln\left(\frac{1}{T}\right)\right)^2 \tag{2.58}$$

The permittivity is given as

$$\varepsilon_r = \frac{\lambda_0^2}{\mu_r}\left(\frac{1}{\lambda_c^2} - \left(\frac{1}{2\pi L} \ln\left(\frac{1}{T}\right)\right)^2\right) \tag{2.59}$$

Equations (2.58) and (2.59) have an infinite number of roots, as the imaginary part of the term $\ln\left(\frac{1}{T}\right)$ is equal to $j(\theta + 2\pi n)$, where $n = 0, \pm 1, \pm 2, \ldots$, the integer of (L/λ_g). Here, n can be attained by two ways.

The first method is the evaluation of group delay. Equation (2.58) is ambivalent because the phase of the transmission coefficient (T) does not deviate on the increment of material length by a multiple of wavelength [18]. This ambiguity can be resolved by considering that the delay through the material is a function of the total length of the material. To resolve phase ambiguity, the ε_r and μ_r solutions are evaluated, and a comparison is made between the calculated and measured group delays to obtain the correct value of n. The calculated group delay can be obtained from the following expression:

$$\tau_{\text{cal}} = L\frac{d}{df}\sqrt{\frac{\varepsilon_r\mu_r}{c^2} - \frac{1}{\lambda_c^2}} = \frac{1}{c^2}\frac{f\varepsilon_r\mu_r + f^2\frac{1}{2}\frac{d(\varepsilon_r\mu_r)}{df}}{\sqrt{\frac{\varepsilon_r\mu_r f^2}{c^2} - \frac{1}{\lambda_c^2}}}L \tag{2.60}$$

The measured group delay, which can be obtained directly from the network analyzer, is expressed as follows:

$$\tau_{\text{meas}} = -\frac{1}{2\pi}\frac{d\varphi}{df} \tag{2.61}$$

The calculated group delay is related to the change in the wavenumber k with respect to the angular frequency. The correct root, $n = k$, is attained when:

$$\tau_{\text{cal}-k} - \tau_{\text{meas}} \approx 0 \tag{2.62}$$

The second method involves an estimation from λ_g using the initial guess values of $\varepsilon_r^*$and μ_r^* for the material. From Eq. (2.58), we have the following:

$$\frac{1}{\Lambda} = j\left(\frac{\gamma}{2\pi}\right) \tag{2.63}$$

where $\gamma = j\frac{2\pi}{\lambda_o}\sqrt{\varepsilon_r^*\mu_r^* - \left(\frac{\lambda_o}{\lambda_c}\right)^2}$

$$\Re_e\left(\frac{1}{\Lambda}\right) = \frac{1}{\lambda g} \tag{2.64}$$

A comparison of Eqs. (2.63) and (2.64) gives λ_g, and thus, the n value can be obtained. Subsequently, the permittivity ε_r and permeability μ_r can be determined from the n value. Alternatively, the permittivity can be determined from Eqs. (2.57) and (2.58), where the n value is not required. However, this method is only valid for permittivity measurements, as this expression assumes a value of $\mu_r = 1$.

From Eq. (2.57),

$$\frac{1}{\Lambda} = \mu_r\frac{(1-\Gamma)}{(1+\Gamma)}\sqrt{\frac{1}{\lambda_0^2} - \frac{1}{\lambda_c^2}} \tag{2.65}$$

By equating this expression with Eq. (2.58), the permittivity can be achieved

$$\frac{1}{\Lambda^2} = \mu_r^2\frac{(1-\Gamma)^2}{(1+\Gamma)^2}\left(\frac{1}{\lambda_0^2} - \frac{1}{\lambda_c^2}\right) = \left(\frac{\varepsilon_r\mu_r}{\lambda_0^2} - \frac{1}{\lambda_c^2}\right) \tag{2.66}$$

$$\varepsilon_r = \mu_r\frac{(1-\Gamma)^2}{(1+\Gamma)^2}\left(1 - \frac{\lambda_0^2}{\lambda_c^2}\right) + \frac{\lambda_0^2}{\lambda_c^2}\frac{1}{\mu_r} \tag{2.67}$$

Here, ε_r is the relative permittivity, μ_r is the relative permeability, L is the material length, ε_r* is the initial guess value for the permittivity, μ_r* is the initial guess value for the permeability, λ_g is the wavelength in the sample, γ is the propagation constant of the material, c is the velocity of light, and f is the frequency.

The NRW conversion technique is fast, non-iterative, and valid for both coaxial lines and waveguides. However, it also has shortcomings as it is not suitable for low-loss materials, and only short samples can be measured. It also shows divergence at certain frequencies corresponding to multiples of one-half wavelength.

2.4.3 NIST Iterative Conversion Method

The NIST iterative method provides an upgraded iterative conversion process for explaining the transmission line equations for permittivity analysis. As known, the NRW method shows divergence at certain frequencies where the sample length is an integer multiple of one-half wavelength in the sample. The scattering parameter $|S_{11}|$ becomes very small at frequencies corresponding to one-half wavelength. Moreover, a small $|S_{11}|$ causes a very large uncertainty in the phase measurement. Therefore, phase errors dominate the results at these frequencies. To overcome this issue, many researchers have resorted to utilizing short samples. However, the use of short small samples reduces the measurement sensitivity, whereas the use of comparatively long samples can minimize the uncertainty in low-loss materials.

Initially, an approximate permittivity guess value is required for the NIST iterative method. It can be obtained from the NRW method or by user input. Then, the result can be obtained using the following expressions.

The reflection coefficient is defined as in Eq. (2.50):

$$\Gamma = \frac{(\gamma_0/\mu_0) - (\gamma/\mu)}{(\gamma_0/\mu_0) + (\gamma/\mu)}$$

The propagation constant (γ_0) in air can be determined using Eq. (2.39):

$$\gamma_0 = j\sqrt{\left(\frac{\omega}{c}\right)^2 - \left(\frac{2\pi}{\lambda_c}\right)^2}$$

Similarly, propagation constant in material is expressed by Eq. (2.38):

$$\gamma = j\sqrt{\frac{\omega^2\mu_r\varepsilon_r}{c^2} - \left(\frac{2\pi}{\lambda_c}\right)^2}$$

where $\varepsilon = \varepsilon_0\varepsilon_r$ and $\mu = \mu_0\mu_r$.

$$c = \frac{1}{\sqrt{\varepsilon_0\mu_0}} \tag{2.68}$$

Substituting Eq. (2.68) into Eq. (2.38) yields the following expression with $\mu_r = 1$:

$$\gamma = j\sqrt{\omega^2\mu_0\varepsilon_r\varepsilon_0 - \left(\frac{2\pi}{\lambda_c}\right)^2} \tag{2.69}$$

From Eq. (2.50), the reflection coefficient can be obtained with $\mu_r = 1$:

$$\Gamma = \frac{(\gamma_0/\mu_0) - (\gamma/\mu)}{(\gamma_0/\mu_0) + (\gamma/\mu)} = \frac{\gamma_0 - \gamma}{\gamma_0 + \gamma} \tag{2.70}$$

The transmission coefficient is given as follows:

$$T = e^{(-\gamma L)} = e^{-jL\left(\sqrt{\omega^2 \mu_0 \varepsilon_r \varepsilon_0 - \left(\frac{2\pi}{\lambda_c}\right)^2}\right)} \tag{2.71}$$

In order to get permittivity values, one out of two expressions needed to be solved:

$$\frac{1}{2}[(S_{12} + S_{21}) + \beta(S_{11} + S_{22})] = \frac{Z(1 - \Gamma^2) + \beta\Gamma(1 - z^2)}{1 - z^2\Gamma^2} \tag{2.72}$$

$$S_{11}S_{22} - S_{21}S_{12} = \left[e^{(-2\gamma_0(L_{\text{air}} - L))}\right] \frac{T^2 - \Gamma^2}{1 - \Gamma^2 T^2} \tag{2.73}$$

The above scattering equations can be solved in several ways, depending on the information available. In measurements where the sample length and reference plane positions are well known, the iterative calculation method yields very stable results for samples of arbitrary length by considering various linear combinations of the scattering equations.

The scattering parameters used in Eq. (2.72) must be transformed from the calibration plane to the sample face using Eqs. (2.47) and (2.48). In Eq. (2.72), β is a constant that varies as a function of three terms: (i) the sample length, (ii) the uncertainty in the S-parameters, and (iii) the loss characteristics of the material. Generally, β is defined as the ratio of the uncertainty in S_{21} to the uncertainty in S_{11}. In low-loss materials, S_{21} is dominant and β can be assumed as zero. In contrast, S_{11} is strong in high-loss materials and β is large. Moreover, the NIST iterative method relies on the initial guess value to yield a superior estimation of the permittivity.

Equation (2.73) is appropriate when the position of the reference plane is unclear. Here, no reference plane transformation is required owing to the use of the relation $L_{\text{air}} = L_1 + L_2 + L$. This approach functions well for both high- and low-loss materials. The uncertainty measurement for this technique provides a general relation for both low- and high-loss materials. The uncertainty decreases as a function of increasing sample length for low-loss materials, whereas high-loss materials show an increase in uncertainty with increasing sample length.

Here, L is the material length, $L_{\text{air}} = L_1 + L_2 + L$ is the length of the air line, ε_r is the relative permittivity, μ_r is the relative permeability, λ_0 is the free space wavelength, λ_c is the cutoff wavelength, c is the velocity of light, and ω is the angular frequency.

In comparison to the NRW method, the NIST iterative method demonstrates smoother, accurate permittivity results with no divergence. This method can be used for both low- and high-loss materials with any arbitrary sample length. However, as a disadvantage, the NIST iterative method can only be used for permittivity measurements and it requires an initial guess value for the permittivity.

2.4.4 New Non-iterative Conversion Method

Boughriet et al. suggested a method to suppress the inaccuracy peaks observed at multiples of one-half wavelength [19]. In non-iterative methods, different formulations of the NRW method have been proposed for the measurement. The approach also uses either all (S_{11}, S_{21}, S_{12}, and S_{22}) or a couple (S_{11} and S_{21}) of the

S-parameters of the MUT to determine the reflection and transmission coefficients. The non-iterative method overcomes the limitations of the NRW method and the NIST iterative method, as no divergence is detected at the frequencies corresponding to the multiples of $\lambda/2$, and an initial guess is not required for the permittivity. Moreover, this method allows fast calculations and has accuracies equivalent to those of the iterative method. A VNA is used to determine the scattering parameters for the measurement. A VNA yields scattering parameters for the measurement.

As known, the reflection coefficient can be expressed by Eq. (2.54)

$$\Gamma = X \pm \sqrt{X^2 - 1}$$

Here, $|\Gamma_1| < 1$ is required for obtaining the correct root.

Similarly, as defined in terms of S-parameters in Eq. (2.55) in the NRW method

$$X = \frac{S_{11}^2 - S_{21}^2 + 1}{2S_{11}}$$

The transmission coefficient can be expressed as follows:

$$T = \frac{S_{11} + S_{21} - G}{1 - (S_{11} + S_{21})G}$$

$$\frac{1}{\Lambda^2} = \left(\frac{\varepsilon_r \mu_r}{\lambda_0^2} - \frac{1}{\lambda_c^2}\right) = -\left(\frac{1}{2\pi L} \ln\left(\frac{1}{T}\right)\right)^2$$

Here, λ_0 is the free space wavelength and λ_c is the cutoff wavelength with

$$\lambda_{og} = \frac{1}{\sqrt{\frac{1}{\lambda_0^2} - \frac{1}{\lambda_c^2}}} \tag{2.74}$$

which denotes the wavelength in an empty cell.

The above equations are the same as those in the NRW method, but an additional intermediate step is included by introducing effective electromagnetic parameters (ε_{eff} and μ_{eff}). These parameters, which assume transverse electromagnetic propagation in the measurement cell, are defined as follows:

$$\mu_{eff} = \frac{\lambda_{og}}{\Lambda} \frac{(1 - \Gamma)}{(1 + \Gamma)} \tag{2.75}$$

$$\varepsilon_{eff} = \frac{\lambda_{og}}{\Lambda} \frac{(1 - \Gamma)}{(1 + \Gamma)} \tag{2.76}$$

$$\mu_r = \mu_{eff} = \frac{1}{\Lambda} \frac{(1 - \Gamma)}{(1 + \Gamma)} \frac{1}{\sqrt{\frac{1}{\lambda_0^2} - \frac{1}{\lambda_c^2}}} \tag{2.77}$$

and

$$\varepsilon_r = \left(1 - \frac{\lambda_0^2}{\lambda_c^2}\right) \varepsilon_{eff} + \frac{\lambda_0^2}{\lambda_c^2} \frac{1}{\mu_{eff}} \tag{2.78}$$

The effective electromagnetic parameters (ε_{eff}, μ_{eff}) can be obtained by using Eq. (2.75) and (2.76), which in place attained from first reflection coefficient (Γ)

and transmission coefficient (T):

$$\Gamma = \frac{\sqrt{\mu_{\text{eff}}/\varepsilon_{\text{eff}}} - 1}{\sqrt{\mu_{\text{eff}}/\varepsilon_{\text{eff}}} + 1} \tag{2.79}$$

$$T = \exp\left(-j\frac{2\pi}{\lambda_{\text{og}}}\sqrt{\mu_{\text{eff}}\varepsilon_{\text{eff}}}\,L\right) \tag{2.80}$$

The electromagnetic parameters (ε_r and μ_r) of the MUT are then obtained using Eqs. (2.77) and (2.78) by equating the propagation constant and impedance with the below expressions:

$$\gamma = \gamma_0\sqrt{\mu_{\text{eff}}\varepsilon_{\text{eff}}} \tag{2.81}$$

$$Z = Z_0\sqrt{\frac{\mu_{\text{eff}}}{\varepsilon_{\text{eff}}}} \tag{2.82}$$

From the above analysis method, a new expression for the permittivity calculation is established for dielectric materials. For dielectric materials ($\mu_r = \mu_{\text{eff}} = 1$), the following expression can be obtained from Eqs. (2.56) and (2.57):

$$\varepsilon_{\text{eff}} = \varepsilon_{\text{eff}}\mu_{\text{eff}} = \frac{\lambda_{\text{0g}}^2}{\Lambda^2} \tag{2.83}$$

The term $(1 - \Gamma/1 + \Gamma)$ in Eqs. (2.75) and (2.76), which is the source of the inaccuracy in the peaks, is eliminated from Eq. (2.83) and thus the inaccuracy is inhibited.

For dielectric materials ($\mu_r = \mu_{\text{eff}} = 1$), the effective permittivity can be expressed more generally as follows:

$$\varepsilon_{\text{eff}} = \varepsilon_{\text{eff}}(\mu_{\text{eff}})^n = \left(\frac{1-\Gamma}{1+\Gamma}\right)^{n-1}\left(\frac{\lambda_{\text{og}}}{\Lambda}\right)^{n+1} \tag{2.84}$$

Here, the exponent n is a real integer and can have a positive or negative value. Processing this general equation with $n = -1$ or 0 comprises the NRW method, whereas $n = 1$ is equivalent to the non-iterative approach. Boughriet et al. who carried out a comprehensive uncertainty analysis, found that as n approaches or reaches 1, the amplitude of the inaccuracy peak decreases or disappears [19]. Moreover, this method can be used for microstrips, coplanar lines, or rectangular waveguide systems. This method also has the advantage of being steady over the entire frequency range for an arbitrary sample length.

2.4.5 Short-Circuit Line (SCL) Method

The SCL measurement method was presented by Robert and Von Hippel as a precise broadband measurement technique. This one-port measurement is applicable to both coaxial lines and waveguides. In this method, only the reflection scattering parameters (S_{11} or S_{22}) of the MUT are used to calculate the reflection coefficient. Moreover, the sample length and position-measured input values are required, and this method is only appropriate for permittivity calculations. Figure 2.6 shows the general setup of a short-circuited line with a sample located in a specified segment.

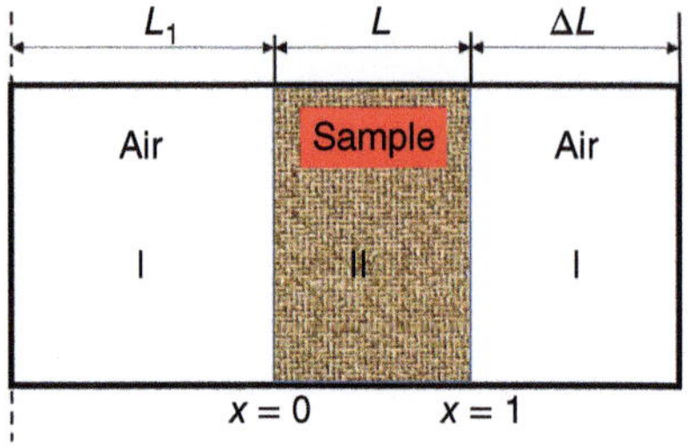

Figure 2.6 Transmission line with a short-circuit termination. Source: Chong Min Koo.

Assuming that only the dominant mode is present in the waveguide, the following equations can be used to express the electric fields in different regions.

$$E_{\mathrm{I}} = \exp(-\gamma_0 x) + C_1 \exp(\gamma_0 x) \tag{2.85}$$

$$E_{\mathrm{II}} = C_2 \exp(-\gamma x) + C_3 \exp(\gamma x) \tag{2.86}$$

$$E_{\mathrm{III}} = C_4 \exp(-\gamma_0 (x - L)) + C_5 \exp(\gamma_0 (x - L)) \tag{2.87}$$

where

$$\gamma = j\sqrt{\frac{\omega^2 \mu_r \varepsilon_r}{c^2} - \left(\frac{2\pi}{\lambda_c}\right)^2} \tag{2.88}$$

$$\gamma_0 = j\sqrt{\left(\frac{\omega}{c}\right)^2 - \left(\frac{2\pi}{\lambda_c}\right)^2} \tag{2.89}$$

Here, γ_0 and γ represent the propagation constants in air and in the MUT, respectively. All the five constants (C_1 to C_5) in Eqs. (2.85)–(2.87) can be obtained through the boundary conditions. The following boundary conditions are imposed at the interfaces ($x = 0$ and $x = L$):

- The electric field is null at the short-circuit position.
- The tangential components of the electric field and magnetic field are continuous at the MUT interface.

$$1 + C_1 = C_2 + C_3 \tag{2.90}$$

$$C_2 \exp(-\gamma L) + C_3 \exp(\gamma L) = C_4 + C_5 \tag{2.91}$$

$$\frac{\gamma_0 \mu}{\gamma \mu_0}[C_1 - 1] = C_3 - C_2 \tag{2.92}$$

$$\frac{\gamma \mu_0}{\gamma_0 \mu}[C_2 \exp(-\gamma L) - C_3 \exp(\gamma L)] = C_4 - C_5 \tag{2.93}$$

$$C_5 = -\delta C_4 \tag{2.94}$$

where

$$\delta = \exp(-2\gamma_0 \Delta L) \tag{2.95}$$

and ΔL is the distance from the short circuit to the MUT. The following equations can be obtained from Eqs. (2.90)–(2.94):

$$C_1 - C_2 - C_3 = -1 \tag{2.96}$$

$$\eta C_2 + \frac{C_3}{\eta} + (\delta - 1)\, C_4 = 0 \tag{2.97}$$

$$C_1 + \beta C_2 - \beta C_3 = 1 \tag{2.98}$$

$$\beta\eta C_2 - \frac{\beta}{\eta} C_3 - C_4(\delta + 1) = 0 \tag{2.99}$$

Here,

$$\beta = \frac{\gamma\mu_0}{\gamma_0\mu} \tag{2.100}$$

$$\eta = \exp(-\gamma L) \tag{2.101}$$

where $1/\beta$ is the effective impedance. The above equations can be expressed in matrix form as follows:

$$\begin{bmatrix} 1 & -1 & -1 & 0 \\ 0 & \eta & \frac{1}{\eta} & \delta - 1 \\ 1 & \beta & -\beta & 0 \\ 0 & \beta\eta & \frac{-\beta}{\eta} & (1+\delta) \end{bmatrix} \begin{bmatrix} C_1 \\ C_2 \\ C_3 \\ C_4 \end{bmatrix} = \begin{bmatrix} -1 \\ 0 \\ 1 \\ 0 \end{bmatrix} \tag{2.102}$$

The permittivity in terms of the reflection coefficient can be determined from Eqs. (2.96)–(2.97), and $\rho = S_{11} = C_1$, with the sample located at a distance ΔL from the short:

$$S_{11} = \frac{-2\beta\delta + ((\delta + 1) + (\delta - 1)\beta^2) \tan \mathrm{h}\gamma L}{2\beta + ((\delta + 1) - (\delta - 1)\beta^2 \tan \mathrm{h}\gamma L)} \tag{2.103}$$

Equation (2.103) can be expressed for hyperbolic functions as follows:

$$S_{11} = \frac{\tan \mathrm{h}\gamma L + \beta \tan \mathrm{h}\gamma_0 \Delta L - \beta(1 + \beta \tan \mathrm{h}\gamma L \tan \mathrm{h}\gamma_0 \Delta L)}{\tan \mathrm{h}\gamma L + \beta \tan \mathrm{h}\gamma_0 \Delta L + \beta(1 + \beta \tan \mathrm{h}\gamma L \tan \mathrm{h}\gamma_0 \Delta L)} \tag{2.104}$$

Equation (2.104) is derived by assuming that the sample plane coincides with the measurement calibration plane. Although this is not the common case, a simple process can transform the reference plane position. To achieve this, a general expression for the reflection coefficient is given as

$$S_{11(\mathrm{trans})} = R_1^2 S_{11} \tag{2.105}$$

Here, $S_{11(\mathrm{trans})}$ denotes the reflection coefficient at the reference plane of calibration:

$$R_1 = \exp(-\gamma_0 L_1) \tag{2.106}$$

and L_1 is the distance from the calibration plane to the sample front face. Moreover, Eq. (2.105) transforms the reflection, and S_{11} can be obtained from Eq. (2.103) or (2.104). The effect of L_1 in Eq. (2.105) can be eliminated by measuring the S_{11} parameter of the empty sample holder.

$$S_{11(\mathrm{empty})} = -\exp(-2\gamma_0[L_1 + \Delta L + L]) = -\exp(-2\gamma_0 L_{\mathrm{air}}) \tag{2.107}$$

Table 2.1 Comparison between the conversion techniques.

S. No.	Conversion methods	S-parameters	Dielectric attributes
1.	Nicholson–Ross–Weir (NRW)	$S_{11}, S_{21}, S_{12}, S_{22}$ or S_{11}, S_{21}	ε_r, μ_r
2.	NIST iterative	$S_{11}, S_{21}, S_{12}, S_{22}$ or S_{11}, S_{21}	$\varepsilon_r, \mu_r = 1$
3.	New non-iterative	$S_{11}, S_{21}, S_{12}, S_{22}$ or S_{11}, S_{21}	$\varepsilon_r, \mu_r = 1$
4.	Short-circuit line (SCL)	S_{11}	ε_r

From Eqs. (2.105) and (2.107)

$$\frac{S_{11(\text{trans})}}{S_{11(\text{empty})}} = -\exp(2\gamma_0[\Delta L + L])S_{11} \tag{2.108}$$

Typically, only one complex parameter, either the permittivity or the permeability, can be measured by the SCL method. Therefore, the experimental analysis provides the reflection coefficient S_{11} at the front face of the MUT, and Eq. (2.104) yields the propagation constant within the MUT, from which the electromagnetic properties can be obtained.

In the arrangement shown in Figure 2.3, a standing wave is formed in the segment between the MUT and the short circuit and in the segment between the calibration plane and the front face of the MUT. Consequently, certain frequencies, depending on the MUT length and other lengths, yield the accurate permittivity, and likewise, other frequencies yield the precise permeability. The position of the short circuit is a region of low electric field and high magnetic field, whereas the position $\lambda/4$ from the short circuit is a region of high electric field and low magnetic field. Thus, for permittivity measurements, the MUT should be placed away from the short-circuit termination, whereas for permeability measurements, the MUT should be placed near the short-circuit termination. A comprehensive analysis can be found elsewhere [14, 20, 21].

All four conversion techniques have been used to obtain the permittivity and permeability from the S-parameters. Table 2.1 provides a summary of the dielectric attributes achieved through the various methods using different sets of S-parameters.

2.5 Summary

An understanding of the EMI shielding mechanisms and scattering parameter conversion theories is essential for designing materials with high-performance SE. The interaction between EMWs and a shield material results in three main shielding mechanisms: reflection, absorption, and multiple reflections. The contribution of each shielding mechanism to the total SE can be altered by varying factors including the material parameters (electrical conductivity, dielectric permittivity, and magnetic permeability), distance from the EMI source, shield thickness, and the frequency. The appropriate selection of both a testing method and a scattering parameter conversion technique is also important for the precise design of materials, as it enhances the reliability of the test results.

References

1 Iqbal, A., Sambyal, P., and Koo, C.M. (2020). 2D MXenes for electromagnetic shielding: a review. *Advanced Functional Materials* 30: 2000883.

2 Ott, H.W. (2011). *Electromagnetic Compatibility Engineering*. Wiley.

3 Kaiser, K.L. (2005). *Electromagnetic Shielding*. CRC Press.

4 Moore, R. (2016). *Electromagnetic Composites Handbook*, 2e. McGraw-Hill Education.

5 Jaroszewski, M., Thomas, S., and Rane, A.V. (2018). *Advanced Materials for Electromagnetic Shielding: Fundamentals, Properties, and Applications*. Wiley.

6 Li, Z., Wang, Z., Lu, W. et al. (2018). Theoretical study of electromagnetic interference shielding of 2D MXenes films. *Metals* 8 (8): 652.

7 Huo, J., Wang, L., and Yu, H. (2009). Polymeric nanocomposites for electromagnetic wave absorption. *Journal of Materials Science* 44 (15): 3917–3927.

8 Dhawan, S., Ohlan, A., and Singh, K. (2011). *Designing of Nano Composites of Conducting Polymers for EMI Shielding*. InTech.

9 Pawar, S.P., Bhingardive, V., Jadhav, A. et al. (2015). An efficient strategy to develop microwave shielding materials with enhanced attenuation constant. *RSC Advances* 5 (109): 89461–89471.

10 Al-Saleh, M.H. and Sundararaj, U. (2009). Electromagnetic interference shielding mechanisms of CNT/polymer composites. *Carbon* 47 (7): 1738–1746.

11 González, M., Pozuelo, J., and Baselga, J. (2018). Electromagnetic shielding materials in GHz range. *The Chemical Record* 18 (7–8): 1000–1009.

12 Baker-Jarvis, J., Geyer, R.G., and Domich, P.D. (1992). A nonlinear least-squares solution with causality constraints applied to transmission line permittivity and permeability determination. *IEEE Transactions on Instrumentation and Measurement* 41 (5): 646–652.

13 Baker-Jarvis, J., Vanzura, E.J., and Kissick, W.A. (1990). Improved technique for determining complex permittivity with the transmission/reflection method. *IEEE Transactions on Microwave Theory and Techniques* 38 (8): 1096–1103.

14 Baker-Jarvis, J. (1990). *Transmission/Reflection and Short-Circuit Line Permittivity Measurements*. National Institute of Standards and Technology Colorado.

15 Nicolson, A. and Ross, G. (1970). Measurement of the intrinsic properties of materials by time-domain techniques. *IEEE Transactions on Instrumentation and Measurement* 19 (4): 377–382.

16 Kerns, D.M. and Beatty, R.W. (1967). *Basic Theory of Waveguide Junctions and Introductory Microwave Network Analysis*, vol. 13. Pergamon.

17 Weir, W.B. (1974). Automatic measurement of complex dielectric constant and permeability at microwave frequencies. *Proceedings of the IEEE* 62 (1): 33–36.

18 Vicente, A.N., Dip, G.M., and Junqueira, C. (2011). The step by step development of NRW method. *2011 SBMO/IEEE MTT-S International Microwave and Optoelectronics Conference (IMOC 2011).* IEEE.

19 Boughriet, A.-H., Legrand, C., and Chapoton, A. (1997). Noniterative stable transmission/reflection method for low-loss material complex permittivity determination. *IEEE Transactions on Microwave Theory and Techniques* 45 (1): 52–57.

20 Baker-Jarvis, J., Janezic, M.D., Grosvenor, J.H. Jr.,, and Geyer, R.G. (1993). *Transmission/Reflection and Short-Circuit Line Methods for Measuring Permittivity and Permeability*. NIST Technical Note 1355 (revised). National Institute of Standards and Technology, U.S. Department of Commerce.

21 Chen, L.-F., Ong, C.K., Neo, C.P. et al. (2004). *Microwave Electronics: Measurement and Materials Characterization.* Wiley.

3

Measurements and Standards

3.1 Introduction

Over the past few decades, huge advances in science and technology have resulted in electronic devices becoming small, portable, and fast. The electrical and electronic systems in both defense and civilian applications work in the low- to high-frequency range. Instruments and testing standards for electromagnetic interference (EMI) shielding measurements have been established by various standardization agencies/organizations. These standards provide permissible limits for radiated electromagnetic waves (EMWs), assessment methods for the attenuation properties of materials, and testing methods for different materials depending on the intended application. The wide array of working frequency bands prevents a sole standard or testing instrument from being applicable over a complete range of systems. Thus, many different standards and instruments are used to obtain reliable measurements of the shielding effectiveness (SE). Test methods such as coaxial transverse electromagnetic (TEM) cell method, rectangular waveguide method, and methods using antennas or electric/magnetic field probes (shielded box test, shielded room test, and free space test) have been employed to achieve accurate results. The purpose of this chapter is to provide an overview of the measurement standards and testing systems that are commonly employed to obtain the scattering parameters used to determine the EMI SE. Moreover, standards such as ASTM ES7-83, ASTM D4935, MIL-STD-285, MIL-G-83528, and IEEE-STD-299, on which these measurement techniques are based, are discussed.

3.2 EMI Shielding Effectiveness (SE) Measurements

Many different types of test methods can be used to measure the EMI SE of a given sample. The selection of an appropriate test process generally depends on the sample type and the intended application.

Two-Dimensional Materials for Electromagnetic Shielding, First Edition.
Chong Min Koo, Pradeep Sambyal, Aamir Iqbal, Faisal Shahzad, and Junpyo Hong.

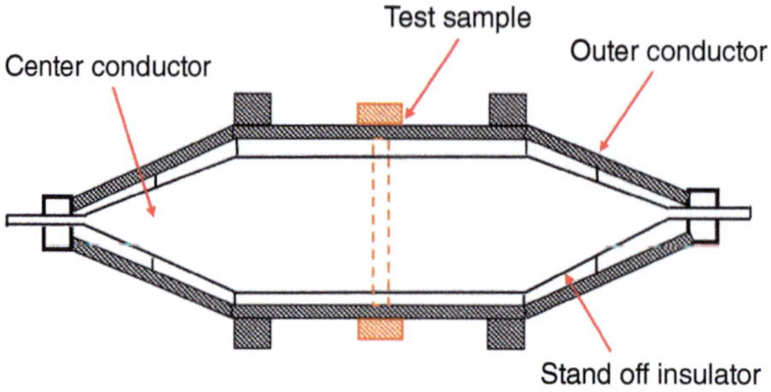

Figure 3.1 Schematic of coaxial transmission line measurements. Source: Chong Min Koo.

3.2.1 Coaxial Transverse Electromagnetic (TEM) Cell Methods

3.2.1.1 Coaxial Transmission Line Method

The coaxial transmission line method is currently the most favored and extensively used technique for SE measurements. This approach overcomes all the limitations of the shielded box method (see Section 3.2.2.2). The significant advantage of this method is that the results or outcomes obtained for a specific sample are analogous at various research facilities. Moreover, the data obtained by coaxial transmission line measurements can be resolved into reflected, absorbed, and transmitted components. In this method, a toroid-shaped sample is used (Figure 3.1). The measurements are carried out at a certain frequency with the help of a signal modulator, tuned amplifier, and crystal detector, or in the sweep mode using a spectrum analyzer receiver and tracking generator. For measurements in the point-to-point mode, the system is set up without a sample holder in the line at a given frequency. Then, the variable attenuator is maximized and the signal level is registered. Subsequently, the sample holder is placed into the line and the attenuator is decreased until the reading is similar to that obtained in the previous step. This signal attenuation is the measure of the SE of the sample. To obtain a spectrum of responses, the entire process is replicated at a series of frequencies. However, this approach for obtaining a spectrum is lengthy and time consuming. In contrast, the sweep mode is quite fast and shows the system response in the form of a single curve on the display screen. In the sweep mode, the generator is substituted with a tracking generator, which is driven by a spectrum analyzer. A dynamic range of 80 dB can be obtained through standard coaxial cables. The coaxial transmission line method is accredited by the American Society for Testing and Materials (ASTM D4935-99) as a standard procedure for measuring the SE of planar samples [1].

3.2.2 Methods Using Antennas or Electric/Magnetic Field Probes

3.2.2.1 Open-Ended Coaxial Probe Method

The open-ended coaxial probe method is a nondestructive testing technique that has been used for many years. During measurements, the probe is immersed in or pressed against the liquid or material under test (MUT) to obtain the reflection coefficient (RC), which yields the permittivity of the sample. This method is suitable for

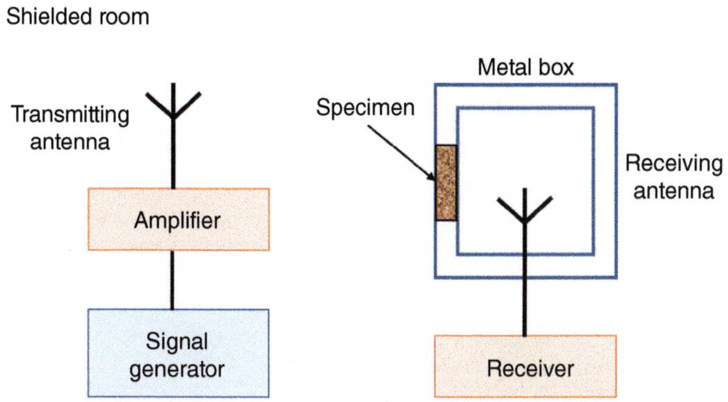

Figure 3.2 Schematic of shielding box SE measurements. Source: Chong Min Koo.

certain applications, such as measurements of biological systems, where the samples cannot be cut or trimmed. Moreover, this method can be used to obtain accurate *in vivo* measurements by placing the probe in close contact with the sample without altering any sample characteristics. Before sample measurement, the vector network analyzer (VNA) probe is calibrated through a direct calibration approach by immersion in reference liquids. The reference liquids should be liquids with known dielectric properties such as water, saline, or methanol. The dielectric properties are obtained by processing the *S*-parameters. As an advantage, this method allows temperature-controlled measurements to be carried out; however, a major drawback is that only reflection measurements can be performed.

3.2.2.2 Shielded Box Method

Figure 3.2 shows a basic diagram of the shielded box method, which is commonly used for relative measurements of test samples of various shield materials. The measurement setup includes a metal box and an electrically tight seam embedded with a sample port in one wall and a receiving antenna. A transmitting antenna outside the box records the signal intensity through the port, both with and without a sample fitted above it. The electromagnetic signals from outside and inside the box are recorded, and the ratio between these signals gives the SE of the sample. It is very difficult to achieve adequate electrical contact between the test sample and the shielding box, which is a major shortcoming of this method. In addition, this method only works well in a limited frequency range of 500 MHz. Furthermore, the results obtained for the same sample in different laboratories or instruments exhibit very poor correlation [2, 3].

3.2.2.3 Shielded Room Method

The shielding room method is more refined than the other test methods and was designed to overcome the drawbacks of the shielded box method. The principle of the shielded room method is similar to that of the shielded box method, except that every component of the measuring system, signal generator, recorder, and transmitting and receiving antenna is kept in a separate isolated enclosure to exclude

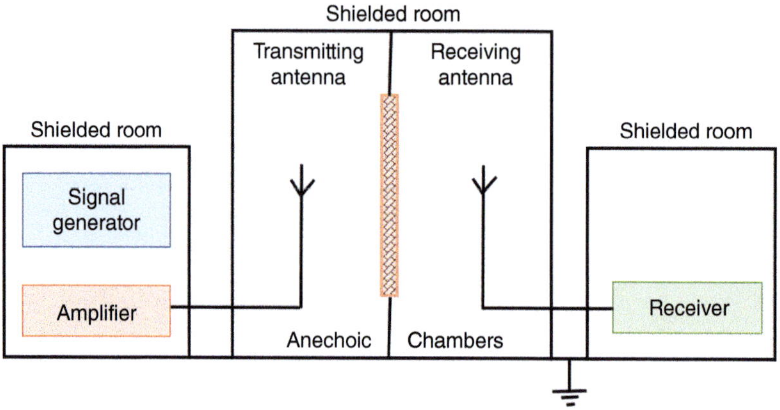

Figure 3.3 Schematic of shielded room SE measurements. Source: Chong Min Koo.

the possibility of interference, as shown in Figure 3.3. Moreover, the antennas are situated in an anechoic chamber and the test sample area is extraordinarily amplified, typically on the order of 2.5 m^2. This technique provides consistent results over an extended frequency range and also delivers better data reproducibility than the shielded box method.

3.2.2.4 Open-Field or Free Space Method

An open-field or free space test assesses both the radiated emission and conductive emission from an electronic system to yield a practical SE. Moreover, this method offers practical conditions for the MUT and evaluates the practical performance of the designed shielding material. In this technique, the device is placed at a distance of 30 m away from the receiving antenna and the radiated emission is recorded (Figure 3.4). In a similar test, the conducted emission transmitted down the power

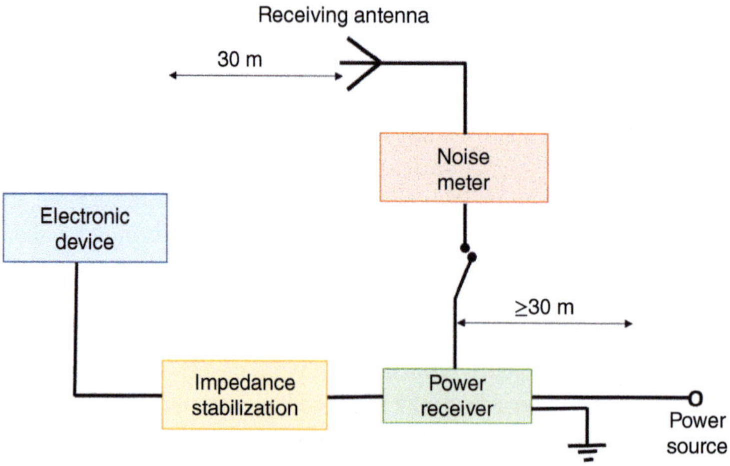

Figure 3.4 Schematic of open-field SE measurements. Source: Chong Min Koo.

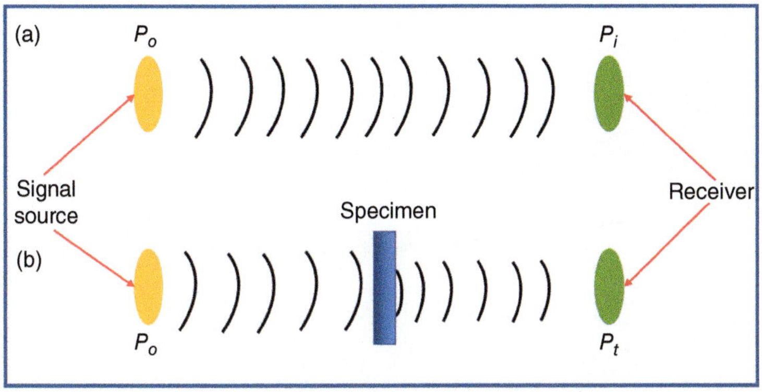

Figure 3.5 Schematic representation of the experimental measurement of SE: (a) reference measurement and (b) load measurement. Source: Chong Min Koo.

line is also recorded. The final result is enumerated using a noise level meter, which indicates the amount of EMI produced [2, 3].

3.3 SE Measurement Systems and Standards

The main objective of SE measurements is to compute the attenuation of an incident wave due to the interaction with the shield material. The basic setup for experimental measurements of the SE is shown in Figure 3.5. The EMI shielding performance of a material is generally assessed by the insertion loss method. In this method, the sample is either in the form of a sheet, pellet, or plate, or the enclosure is made of the shield material. During the measurement, a signal emitted by the source is measured by the receiving antenna with and without the test sample. The SE is obtained from the logarithm of the ratio of the received powers without (P_i) and with the sample (P_t) as follows:

$$\mathrm{SE} = 10 \log \frac{P_i}{P_t} \tag{3.1}$$

The various measurement techniques and standards can be categorized according to the shape, thickness, and morphology of the test sample [1]. SE measurement can be carried out using radiated measuring methods, free space methods, or other techniques. Some characteristic test techniques for EMI SE are described in the subsequent Section 3.4.

3.3.1 SE Calculations Using Experimental Scattering Parameters

EMI SE measurements of test samples are carried out by placing the sample between the flanges connected to the waveguides of a VNA. The scattering parameters are obtained as the VNA emits a signal from one port to another. Moreover, the scattering parameters define the two-port circuit reflection coefficients and transmission coefficients. The reflection coefficients are denoted as S_{11} and S_{22},

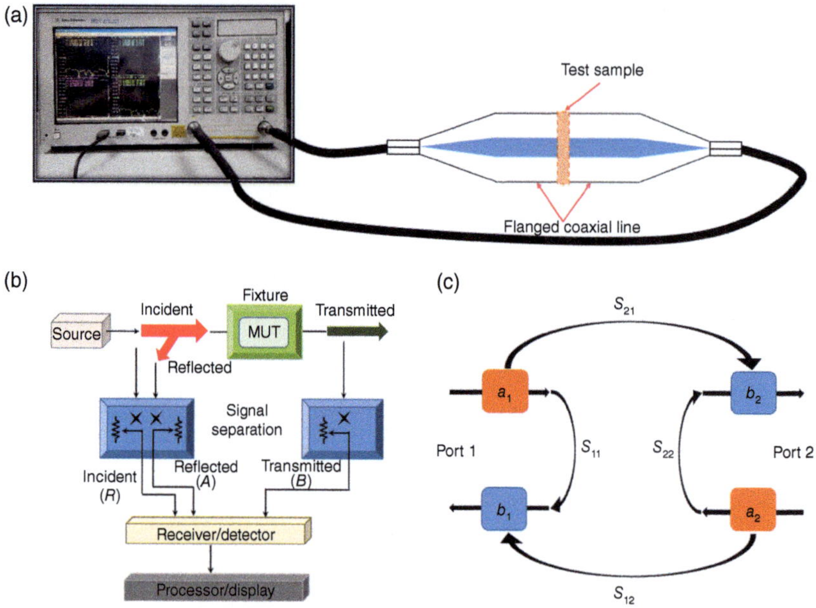

Figure 3.6 (a) Basic setup for SE measurements, (b) internal block diagram of a two-port VNA, and (c) signal paths in a two-port VNA, as described using the scattering parameters. Source: Chong Min Koo.

whereas the transmission coefficients are denoted as S_{12} and S_{21}. Figure 3.6a shows the basic setup for SE measurements consisting of a VNA and a coaxial holder. A passive two-port VNA measurement can provide insights into the scattering parameters, as shown in Figure 3.6b. When an EMW passes from the two ports through the coupled transmission line, it will experience scattering so that it both advances through the ports and reflects back toward the source, as depicted in Figure 3.6c. The S-parameters define the correlation between the incident, reflected, and transmitted waves [4].

In the forward direction of wave propagation, the incident, transmitted, and reflected waves are denoted as a_1, b_2, and b_1, respectively. In the backward direction, the incident, transmitted, and reflected waves are denoted as a_2, b_1, and b_2, respectively, as shown in Figure 3.6c. The reflection coefficients (S_{11} and S_{22}) and transmission coefficients (S_{12} and S_{21}) can be obtained using the following expression [5, 6]:

$$S_{11} = \left.\frac{b_1}{a_1}\right|_{a_2=0} \quad S_{21} = \left.\frac{b_2}{a_1}\right|_{a_2=0} \quad S_{12} = \left.\frac{b_1}{a_2}\right|_{a_1=0} \quad S_{22} = \left.\frac{b_2}{a_2}\right|_{a_1=0}$$

The S-parameters provide relationships for the transmittance (T), reflectance (R), and absorbance (A) of the shielding materials as follows:

$$T = \left|\frac{E_T}{E_I}\right|^2 = S_{12}^2 = S_{21}^2 \tag{3.2}$$

$$R = \left|\frac{E_T}{E_I}\right|^2 = S_{11}^2 = S_{22}^2 \tag{3.3}$$

$$A = 1 - R - T \tag{3.4}$$

When the shielding due to absorption loss (SE_A) is greater than 10 dB, the shielding due to multiple reflections is considered negligible; therefore, the total shielding is considered as the sum of SE_R and SE_A. The term $(1 - R)$ defines the intensity of an EMW inside the shielding material after the primary reflection, which yields the effective absorbance (A_{eff}) as follows:

$$A_{\text{eff}} = 1 - R - T/1 - R \tag{3.5}$$

The shielding due to reflection loss (SE_R) and absorption loss (SE_A) are given as

$$SE_R = -10\log_{10}(1 - R) \tag{3.6}$$

$$SE_A = -10\log_{10}(1 - A_{\text{eff}}) \tag{3.7}$$

Therefore, the reflection loss and absorption loss components of the total shielding can be obtained from the known reflectance (R) and transmittance (T) values [7].

3.4 Methods and Standards

3.4.1 Coaxial TEM Cell Methods

TEM cells, which consist of an extended coaxial transmission line with conical ends connected to the normal coaxial line, are used as sample holders in EMI SE measurements. Moreover, TEM cells simulate far-field conditions for the measurements. TEM cell methods are widely accepted and used and are the foundation of the American standards ASTM ES7-83 and ASTM D4935-99. TEM cells are generally designed with circular or rectangular cross sections and an impedance of 50 Ω. The frequency range in which a TEM can be used generally depends on its geometry, with larger geometries lowering the usable upper frequency. TEM cell measurements provide a simple way to measure the radiated fields emanating from the MUT and have been extensively used for component evaluations (liquid crystal displays [LCDs] and power supply modules), field probe calibrations, and SE measurements. Figure 3.7 depicts the various measurement kits typically used for laboratory-based EMI measurements. Measurement kit selection is based on parameters such as the geometry or size of the sample, the physical attributes of the sample, and the frequency range of the measurement. Even the ASTM ES7 and ASTM D4935 standards provide one coaxial measurement kit with two assemblies for measurements in the frequency ranges of 10 kHz to 3 GHz and 45 MHz to 3 GHz. For thin films or plate-type samples, measurements can be carried out using a parallel plate setup, as shown in Figure 3.7, which works in the frequency range of 30 MHz to 1.5 GHz, based on ASTM D4935-1. The coaxial airline measurement kit provides a large frequency band (300 kHz to 18 GHz) for EMI SE measurements. Bulk samples or thin films can also be measured

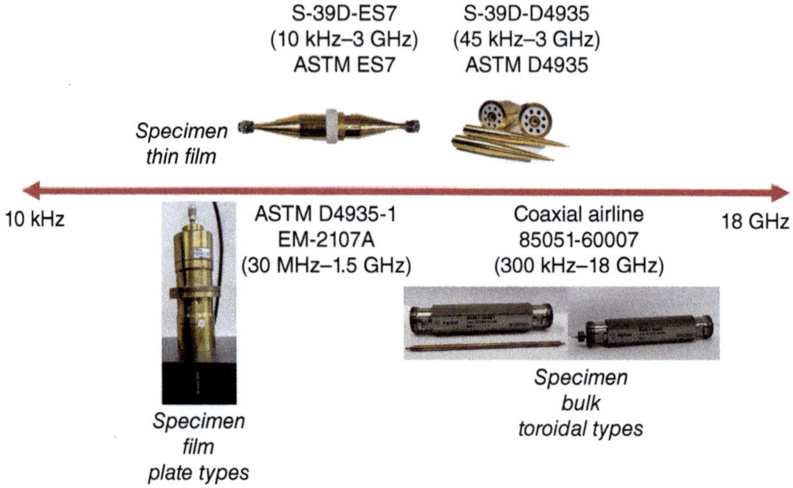

Figure 3.7 Standards and general measurement kits for TEM cell measurements. Source: Chong Min Koo.

in both the X-band (8.2–12.4 GHz) and Ku-band (12.4–18 GHz) through waveguides. Thus, measurements can be carried out for different samples from films to bulk powders in various frequency ranges.

3.4.1.1 ASTM ES7-83 Method

The ASTM ES7-83 method is based on a circular coaxial transmission line holder with a continuous conductor, as shown in Figure 3.8. The test fixture consists of an inner and an outer conductor. The inner conductor has a broad diameter and narrows toward the ends to connect with the center pin of the connector. To maintain an impedance of 50 Ω, the size of the outer conductor is extended to appropriate dimensions. Longitudinal slits are located in the middle sections to hold a disk-shaped test sample. A continuous connection is obtained by assembling both the outer and inner conductors. Only one TEM propagation mode is allowed in the TEM cell in the

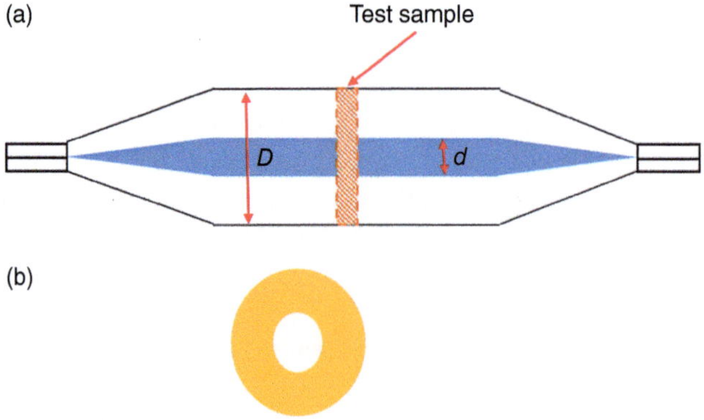

Figure 3.8 (a) ASTM ES7 test cell and (b) washer-shaped sample. Source: Chong Min Koo.

frequency range of 1 MHz to 1.8 GHz. Advancement and modification of the present method has yielded a low contact resistance and a frequency range of 0 Hz to 5 GHz [8–10]. To prevent perturbation, the upper frequency should not exceed the cutoff frequency (f_{max}):

$$f_{max} < \frac{c}{\frac{\pi}{2}(D + d)} \tag{3.8}$$

Here, D is the inner diameter of the outer conductor, d is the diameter of the inner conductor, and c is the speed of light. The measurement can be carried out using a signal source and receiver, such as an oscilloscope, spectrum analyzer, or VNA. Silver paint is used to form electrical contacts between the inner and outer conductors, and the edges of the material should also be painted for proper measurements. The insertion loss method is used to quantify the EMI SE of a test sample by employing Eq. (3.1). The transmission performance of this device is remarkable. However, a major shortcoming of this method is that the test fixture does not provide significant values for test samples such as coatings or materials containing embedded conductive layers or high surface resistivity [1, 6, 11].

3.4.1.2 ASTM D4935 Method

The ASTM D4935 method, which was developed and issued in 1989, is the most suitable technique for measuring the SE of thin, flat samples and composite materials. Moreover, it can be used at higher frequencies for plane-wave SE measurements. The setup is composed of a coaxial test cell with a flanged outer conductor and a discontinuous inner conductor (Figure 3.9). The primary experimental setup consists of two components: a VNA and a sample holder. The VNA can easily measure the incident, reflected, and transmitted powers. The sample holder consists of an enlarged coaxial transmission line (with an external diameter of 76 mm and an internal diameter of 33 mm) with tapered sections and notched grooves, which allows an impedance of 50 Ω to be maintained throughout the length of the sample holder. The insertion loss method is used to determine the SE of the test sample. For

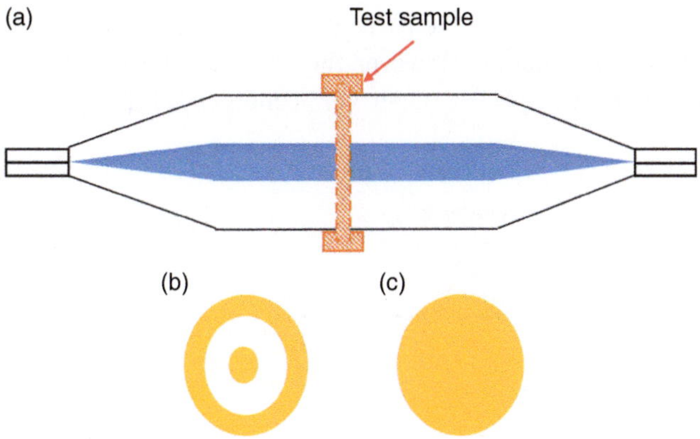

Figure 3.9 (a) ASTM D4935 test cell, (b) reference sample, and (c) load sample. Source: Chong Min Koo.

the measurement, a reference sample with a diameter similar to that of the outer conductor flange is placed in the fixture, and the incident power is recorded. The load measurement is obtained using a sample disk with a diameter equal to that of the outer conductor flange, and the transmitted power is measured. The SE of the test sample can then be obtained using Eq. (3.1).

This measurement technique is valid in the frequency range of 30 MHz to 1.5 GHz; however, the upper frequency should not exceed the cutoff frequency. Recent developments have extended the frequency range up to 18 GHz [10, 12]. This technique provides more precise results than the ASTM ES7 standard for various materials, such as metals, conductive plastics, and samples with high surface resistivities. However, the instrument required for this method is similar to that used for the ASTM-ES-7 standard. There should not be any connections between the cell components, especially the outer conductor, during the measurement process, as the results are adversely affected by the contact resistance generated by such contacts. To overcome this issue, nonconductive screws or supports are used to fasten the flanges to the sample [1, 13].

3.4.1.3 TEM-t Cell Method

The TEM-t cell is generally used to measure low- or medium-conductivity thin composite materials composed of an insulating matrix packed with conductive fillers or with conductive films/layers deposited on it. The physical dimensions of the TEM-t cell are similar to those of the ASTM D4935 test cell, and the two-part assembly has a rectangular cross section, as shown in Figure 3.10. Each part has a conducting outer section with tapered and rectangular regions with large flanges for high capacitive coupling. Moreover, the flat inner conductor is 1 mm shorter than the plane of the flanges to avoid direct contact with the MUT. The design of the TEM-t cell offers a capacitive coupling mechanism that is independent of the sample surface conductivity [14–16]. The measurement procedure is similar to those for the above-described methods. Before the measurement, a reference measurement is carried out by directly connecting both parts of the cell, whereas the loaded measurement is performed by placing the sample between the parts. It is known that higher order modes arise from asymmetrical excitation. However, the TEM-t cell is symmetrical and can be used up to a frequency of 12 GHz under TEM conditions [15].

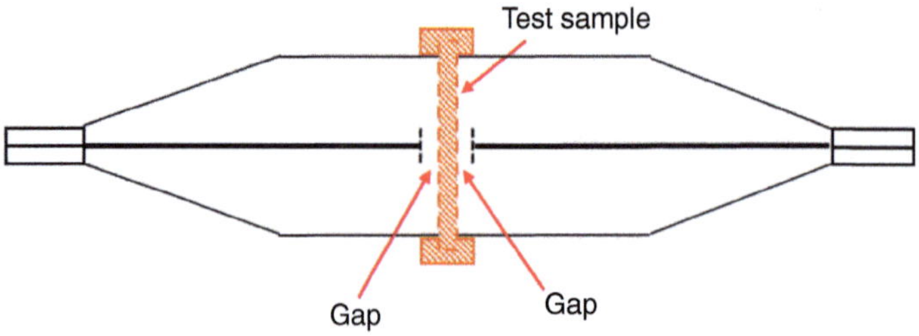

Figure 3.10 TEM-t cell. Source: Chong Min Koo.

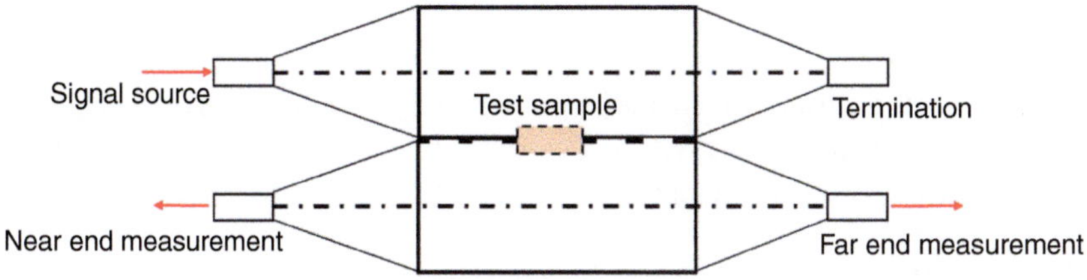

Figure 3.11 Dual TEM cell. Source: Chong Min Koo.

3.4.1.4 Dual TEM Cell Method

The dual TEM cell method is relatively an economical method. A dual TEM cell fixture can easily isolate the coupled electric and magnetic fields. The distinctive feature of this method is that near-field SE measurements can be achieved, including the SE of both the E-field and H-field. The dual TEM cell is composed of two TEM cells connected through a shared aperture where the test specimen is installed. One of the TEM cells is connected to the signal source on one end and the other end is terminated, as depicted in Figure 3.11. The first TEM cell serves as a driving cell and transmits power through the aperture to the second cell, which serves as a receiver. The second cell yields two outputs, allowing the electric field and magnetic field coupling to be assessed. Consequently, the following expressions yield the electric and magnetic SE:

$$\mathrm{SE}^{e} = 20\ \log\left|\frac{\sum i}{\sum t}\right| \tag{3.9}$$

$$\mathrm{SE}^{m} = 20\ \log\left|\frac{\Delta i}{\Delta t}\right| \tag{3.10}$$

where $\sum i$ and $\sum t$ represent the sums of the signals recorded at the forward and backward ports with unloaded and loaded apertures, respectively. Similarly, Δi and Δt are the differences between the signals measured at the same ports. To obtain the signal sums and differences in the receiving cell, a VNA can be used. In this method, the electric and magnetic field shielding can be instantaneously determined by measuring the received power at both ends of the secondary TEM cell. This cell is serviceable in the frequency range of 1–200 MHz and the dynamic range is approximately 60 dB. One shortcoming of this technique is that the electric field polarization is normal to the test sample [1, 3, 11, 13].

3.4.1.5 Split TEM Cell

The split TEM cell, also known as a rectangular split transmission line holder, is another type of TEM cell. The split TEM cell is composed of an ordinary TEM cell in two halves. The SE of the test specimen is obtained by measuring the attenuation through the empty cell with both halves joined and the attenuation through the shorted cell with a sample between the joined halves. There should be good contact between the inner and outer conductors and the sample on both sides so that the cell can be shorted by the sample. The magnetic SE can be measured by modifying the receiving half of the TEM cell by coupling a loop antenna to a box with a 90° angle reflector on one wall. The loop antenna is installed through the reflector such that 75% of the loop is inside the box and 25% outside. During the measurement, the wall with the reflector and a quarter of the loop antenna is joined with half of the TEM cell with the sample in the middle. The serviceable frequency range for this technique is 1 MHz to 1 GHz, and the dynamic range is approximately 70–80 dB [3, 13].

3.4.1.6 Apertured TEM Cell in a Reverberating Chamber

In this technique, the field source is a reverberating chamber, which consists of a shielded enclosure with a paddle that moves in a continuous manner or in small discreet steps. As there is some deviation in the paddle position, the field in the chamber exhibits an average wave impedance value close to that in free space. This technique is more suitable for higher frequencies. An apertured TEM cell, which acts as a receiver, and a transmitting antenna are located inside the chamber. The positions of the transmitting antenna and the TEM cell are not important, but the antenna should not be facing toward the TEM cell and the TEM cell should not be installed near the wall or other reflecting materials. The TEM cell is an expanded unit of a rectangular coaxial transmission line. The sample is placed on the aperture of the TEM cell. The TEM cell coupled to the antenna and receiving instrument generates an electromagnetic field, which is used to compute the leakage through the test sample. The working frequency for this method is 200 MHz to 1 GHz, and the dynamic range is approximately 100 dB. This method requires very long measurement times at frequencies below 1 GHz because the paddle rotates in discrete steps, whereas above 1 GHz, the paddle rotates continuously, resulting in shorter measurement times [1, 3, 6].

3.4.2 Rectangular Waveguide Method

In this method, waves are propagated through a conducting cylinder with a rectangular cross section. The electromagnetic field is confined and guided by the conducting wall of the waveguide. Rectangular waveguides, which are characterized by their physical dimensions, i.e. length and breadth, support both transverse electric (TE) and transverse magnetic (TM) propagating modes but not TEM modes. Rectangular waveguides are generally used to transmit signals in critical high-frequency applications between 1 and 220 GHz. Each waveguide has a cutoff frequency, below which no signal can propagate along the waveguide. Figure 3.12 shows the different

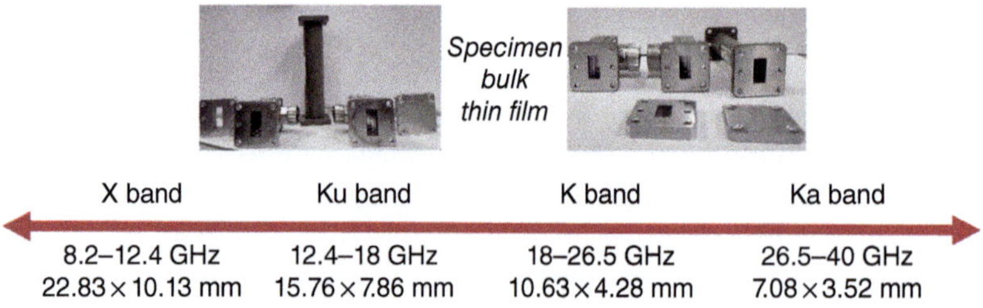

Figure 3.12 Waveguide cells. Source: Chong Min Koo.

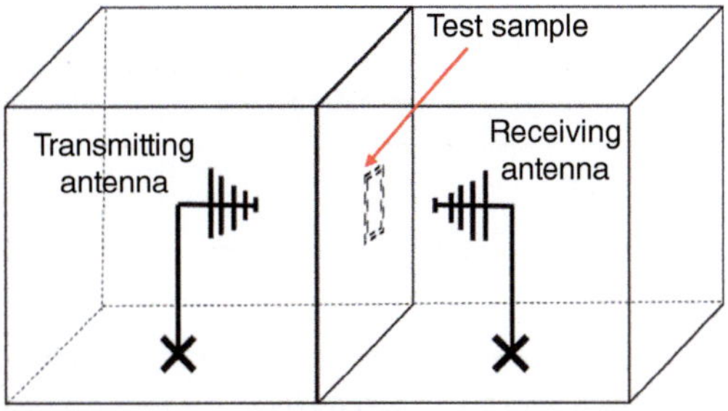

Figure 3.13 Schematic of SE measurements according to the MIL-STD-285 standard. Source: Chong Min Koo.

rectangular waveguides used for measurements in different frequency bands. This method is widely used in research laboratories and industries, owing to its simple handling and easy measurement process. Another advantage of rectangular waveguides is that they conduct microwave energy with lower losses than coaxial cables. Thus, rectangular waveguides are used in radar, microwave communication, and other high-frequency applications.

3.4.3 Methods Using Antennas or Electric/Magnetic Field Probes

3.4.3.1 Testing Methods Based on MIL-STD-285

The MIL-STD-285 standard acts as a basis for SE measurements. The measurement setup includes two shielded rooms with one shared wall, as depicted in Figure 3.13, and the test sample is placed in an opening in the wall. The transmitting and receiving antennas are installed in each room, facing each other at a fixed distance. In the measurement, a constant power is transmitted without and with the test sample, and the transferred power is measured by the receiver to obtain the SE. The MIL-STD-285 method is generally used to characterize the SE performance of conductive composites, thin films, and conductive gaskets. The major disadvantage of this method is poor repeatability arising from the uneven placement of the antennas and wave reflections inside the shielded rooms. An upgraded form of the MIL-STD-285 standard has been designed to overcome the reflection issue by using absorbing material in the shielded rooms. A new improved version, such as MIL-G-83528, can be found in the IEEE-STD-299 standard, which is applicable for the frequency range of few MHz to 18 GHz [11, 17].

Major infrastructures such as residential buildings, military stations, subways, and aviation sites are exposed to high-power electromagnetics (HPEM). As HPEMs can cause great turmoil due to server failures, HPEM protection standards are a matter of national security, and most nations have not disclosed test methods and standards. To avert destruction by HPEM attacks, special installations or facilities are needed. Such installations and facilities have typically been constructed, evaluated, and maintained based on the MIL-STD-188-125 standard, which is the only US military standard disclosed in the open literature [18]. Three basic standards are used for SE testing: MIL-STD-188-125 [19] for assessing C4I (command, control, communications, computers, and intelligence) military facilities, IEEE-STD-299 [20] for evaluating general electromagnetic shielding rooms, and IEC 61000-4-23 [21]

Table 3.1 Comparison of SE test standards.

	IEC 61000-4-23	MIL-STD-188-125-1	IEEE-STD-299
Issue year	2000	2005	2006
Application field	Private facilities (EMP)	Military facilities high-altitude electromagnetic pulse (HEMP)	Private facilities (electromagnetic compatibility [EMC])
Test frequency	15–500 kHz	10–100 kHz	9 kHz to 300 MHz
	300–500 kHz	100 kHz to 1 MHz	300–600 MHz
	1–20 MHz	1–10 MHz	600 MHz to 1 GHz
	150–200 MHz	10–100 MHz	1–2 GHz
		100 MHz to 1 GHz	2–4 GHz
			4–8 GHz
			8–18 GHz
Transmission and receiving antenna distance	Loop antenna Transmission: 0.955 m Receiving: 5–60 cm	Loop/bi-conical/log-periodic (LP) antenna Transmission: 2.05 m Receiving: 1.0 m	Loop antenna Transmission: 0.3 m Receiving: 0.3 m
	Dipole antenna Transmission: 4.7 m Receiving: 5–60 cm		Bi-conical/dipole/horn antenna Transmission: 1.7 m Receiving: 0.3 m
Transmission antenna location	Outside the facility	Outside the facility	Outside the facility
Unit test area	2.5 m × 2.5 m	3.05 m × 3.05 m	2.6 m × 1.5 m

Source: Radasky and Hoad [21]; and Seo et al. [23].

for determining the protection against electromagnetic pulses (EMPs) in non-combatant installations [18, 22, 23]. These standards are applicable to different test environments and parameters such as the distance between transmitting and receiving antennas, test frequency, and unit test area (Table 3.1) [21, 23].

3.4.3.2 Modified Radiation Method Based on MIL-G-83528

The MIL-G-83528 method is a modified version of the MIL-STD-285 standard for SE testing using the radiated energy approach. MIL-STD-285 specifies that the wave produced by the transmitting antenna is focused at the discontinuities or joints between the test specimen and shared wall in the shielded room, and the maximum field strength radiating from the joints is recorded by the positioning of the receiving antenna. In MIL-G-83528 measurements, the EMW is directed at the center of a large cover plate (28 in. × 28 in.), and the receiving antenna is placed just behind the plate, as depicted in Figure 3.14. After measuring an open reference, a

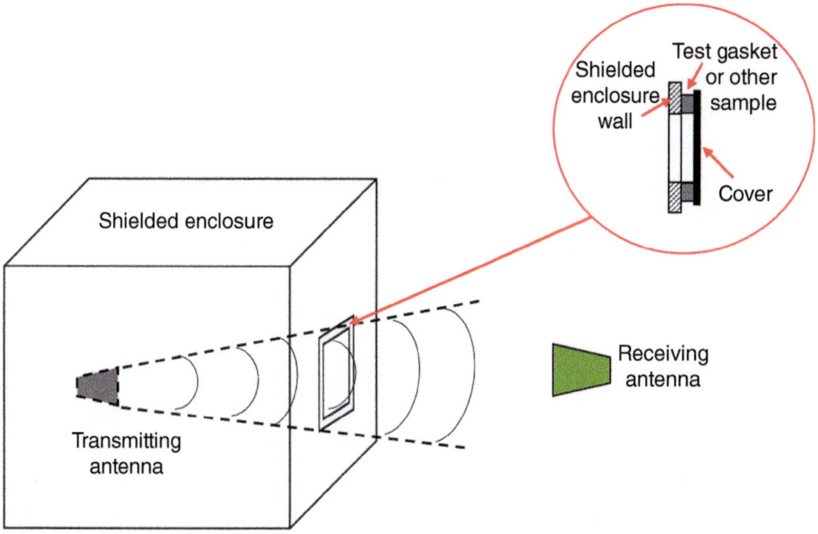

Figure 3.14 MIL-G-83528-based modified radiation testing method. Source: Chong Min Koo.

closed reference is obtained with all the equipment in the same positions and the test specimen installed. The transmitting antenna is connected to a radio frequency amplifier and an appropriate signal generator, whereas the receiving antenna is coupled to a spectrum analyzer. A plastic spacer and compression plate hold the test specimen in place. The compression plate applies an even load and grips the specimen. The SE of the test sample is obtained from the difference between the measured power levels for the open and closed references [3, 11].

3.4.3.3 Dual Mode-Stirred Chamber

A dual mode-stirred chamber was designed and developed by the National Institute of Standards and Technology (NIST, USA), in which two chambers share a common wall with an aperture for installing the test sample. Figure 3.15 shows the

Figure 3.15 Dual mode-stirred chamber facility. Source: Chong Min Koo.

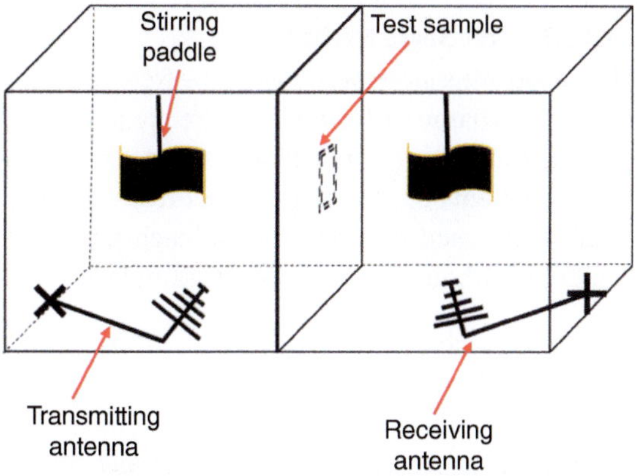

shielded mode-stirred chambers, in which mode stirring creates electromagnetic fields with a large number of modes within the enclosure. Any mechanical paddle wheel or highly conductive structure that can revolve in a stepwise or continuous manner can be used for mode stirring. The signal source can be modulated to create frequency mode stirring. The introduction of a small power amplifier can yield a high-amplitude multimode electromagnetic field in the chamber. Mode tuning can also be carried out by implementing a stirring mechanism between the measurements so that various measurements are obtained in different electromagnetic environments. Moreover, measurements can be performed more easily by allowing the stirring mechanism to continuously change the electromagnetic environment in the enclosure. The position of the antennas is not critical in this measurement method. The major limitation of this system is that very expensive equipment is required to control the mode stirring. The low-frequency limit for this technique (500 MHz) depends on the size of the mode-stirred chamber. The shortest distance in the chamber is assumed to be seven times the wavelength of the lowest frequency; therefore, a distance of more than 4 m is required for 500 MHz. The various ways to derive or obtain the lowest useful frequency provide different lower frequency limits and field uniformity measurements in the completed chamber, which is considered the ideal way to obtain exact realistic values [6, 11].

3.4.3.4 IEEE-STD-299

The IEEE-STD-299 standard was introduced in the late 1960s as a comprehensive measurement method for high-performance shielding enclosures, and many modifications have since been implemented. IEEE-STD-299, which outperforms and replaces MIL-STD-285, provides a uniform method for evaluating the SE of enclosures of at least 2 m in the range of 9 kHz to 100 GHz (Figure 3.16). IEEE-STD-299 can be used for measurements in both the low- and high-frequency regions and to test enclosures made of aluminum, steel, copper, and hybrid composites. A reference measurement is required before the test sample measurement, and the accuracy of the outcome depends on the selection of appropriate antennas for different frequencies. Moreover, different procedures should be used depending on the power passing through the transmitting antenna to the receiving antenna [13, 17].

3.4.3.5 Free Space Methods

Free Space Measurement Method The free space measurement method (Figure 3.17a) is used for large test samples at frequencies up to the gigahertz range. Moreover, this method is applicable to a wide class of shielding materials that require in situ measurements or electrical contact-free measurements. The SE of a test specimen can be obtained from the field at each frequency that is received by the receiver antenna without and with the test sample as follows:

$$\mathrm{SE} = 20\log\frac{E_i}{E_t}$$

Free Space Reflection Measurement Method Figure 3.17b shows the instrumental setup for the free space reflection measurement technique. This technique allows

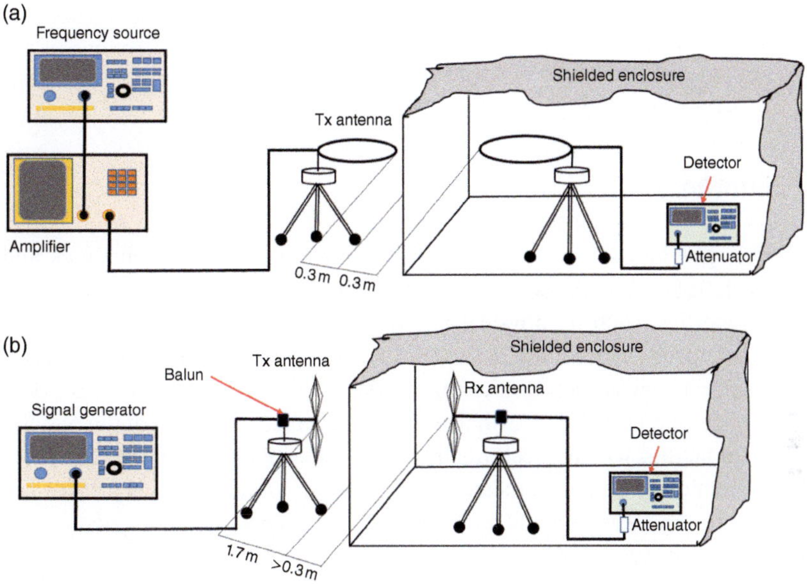

Figure 3.16 Measurement setup for IEEE-STD-299: (a) low-frequency range and (b) resonant and high-frequency range. Source: Chong Min Koo.

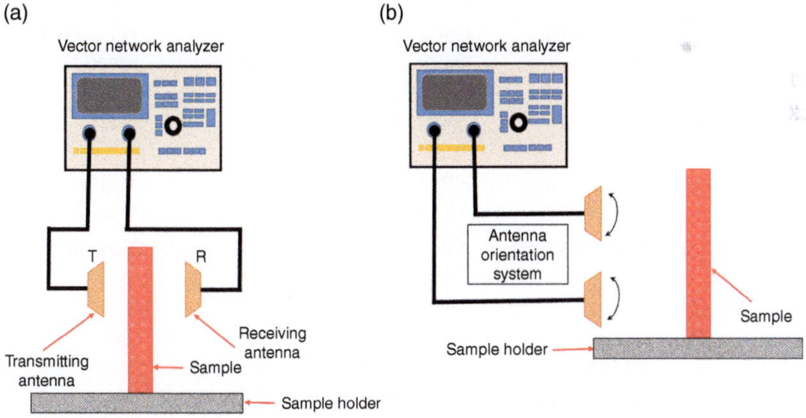

Figure 3.17 (a) Free space transmission measurement method and (b) free space reflection measurement method. Source: Chong Min Koo.

the incidence angle and angular reflectivity of the test specimen to be determined owing to the alignment between the antenna systems. The reflection coefficient is defined as $\mathrm{RC} = 20\log(E_r/E_i)$, where E_i is the incident electric field and E_r is the electric field reflected by the test specimen. The reflectivity of the sample can be obtained from the backscattering coefficient (BC), which is expressed as $\mathrm{BC} = 20\log(E_S/E_{\mathrm{MP}})$, where E_S and E_{MP} are the electric fields received by antenna R for the sample and a metallic reference plate, respectively. A single antenna coupled with a broadband high-directivity directional coupler can be employed for reflection measurements in the case of waves at normal incidence. The main complications of this technique include the separation of the preferred electric field from the perturbing fields and the generation of unwanted reflections in the measurement space [3].

3.5 Summary

This chapter provides a broad discussion of various measurement methods and standard techniques. The measurement method is generally selected based on the material properties, such as physical and dielectric attributes and the required measurement speed or accuracy. To obtain accurate results, it is very important to use a method that is appropriate for the MUT, as specific methods are only applicable to certain materials. Furthermore, the accuracy and speed of the method are important factors, which affect the time required for S-parameter measurement and the conversion of the S-parameters. Comprehensive information is provided in Table 3.2 about the measurement methods and conversion techniques that can be used for various materials [24].

Table 3.2 Suitable measurement methods and conversion techniques for various materials.

Materials/length/ magnetic properties	Measurement methods	Conversion methods	S-parameters	Comprehensive speed
Lossy solids/short/ nonmagnetic	Transmission line	Nicholson–Ross–Weir (NRW)	S_{11}, S_{21}	Fast
Lossy solids/short/ magnetic	Transmission line	NRW	S_{11}, S_{21}	Fast
Low lossy solids/long/ nonmagnetic	Transmission line	NIST iterative/new non-iterative	S_{11}, S_{21}	Slow/fast
Biological sample	Open-ended coaxial probe	Rational function model (RFM)	S_{11}	Fast
Liquids	Open-ended coaxial probe	RFM/reference liquids	S_{11}	Fast
Semisolids	Open-ended coaxial probe	RFM	S_{11}	Fast
High-temperature solids/large/flat/ nonmagnetic	Free space	NIST iterative/new non-iterative	S_{11}, S_{21}	Slow/fast
High-temperature solids/large/flat/ nonmagnetic	Free space	NRW	S_{11}, S_{21}	Fast

References

1 Morari, C. and Balan, I. (2015). Methods for determining shielding effectiveness of materials. *Electrotehnica, Electronica, Automatica* 63 (2): 126.

2 Geetha, S., Kumar, K.K.S., Rao, C.R.K. et al. (2009). EMI shielding: methods and materials – a review. *Journal of Applied Polymer Science* 112 (4): 2073–2086.

3 Jaroszewski, M., Thomas, S., and Rane, A.V. (2018). *Advanced Materials for Electromagnetic Shielding: Fundamentals, Properties, and Applications*. Wiley.

4 Saini, P. and Arora, M. (2012). Microwave absorption and EMI shielding behavior of nanocomposites based on intrinsically conducting polymers, graphene and carbon nanotubes. *New Polymers for Special Applications* 3: 73–112.

5 Langdon, H.S. and Luebbers, R. (1997). Efficient FDTD calculation of multi-port S parameters for microstrip and stripline circuits. *IEEE Antennas and Propagation Society International Symposium 1997. Digest*. IEEE.

6 Lundgren, U. (2004). *Characterization of Components and Materials for EMC Barriers*. Luleå tekniska universitet.

7 Sambyal, P. et al. (2019). FeSiAl/metal core shell hybrid composite with high-performance electromagnetic interference shielding. *Composites Science and Technology* 172: 66–73.

8 Dřínovský, J. and Kejik, Z. (2009). Electromagnetic shielding efficiency measurement of composite materials. *Measurement Science Review* 9 (4): 109–112.

9 Desideri, D. and Maschio, A. (2012). A new version of coaxial holder with continuous conductor for tests on planar films. *International Journal of Applied Electromagnetics and Mechanics* 39 (1–4): 189–194.

10 Tamburrano, A. et al. (2014). Coaxial waveguide methods for shielding effectiveness measurement of planar materials up to 18 GHz. *IEEE Transactions on Electromagnetic Compatibility* 56 (6): 1386–1395.

11 Tong, X.C. (2016). *Advanced Materials and Design for Electromagnetic Interference Shielding*. CRC Press.

12 Sarto, M.S. and Tamburrano, A. (2006). Innovative test method for the shielding effectiveness measurement of conductive thin films in a wide frequency range. *IEEE Transactions on Electromagnetic Compatibility* 48 (2): 331–341.

13 Soyaslan, D.D. (2013). Investigation of test instruments for EM shielding effectiveness of conductive fabrics and their composites. *Journal of Safety Engineering* 2 (2): 39–44.

14 Catrysse, J., Delesie, M., and Steenbakkers, W. (1992). The influence of the test fixture on shielding effectiveness measurements. *IEEE Transactions on Electromagnetic Compatibility* 34 (3): 348–351.

15 Catrysse, J. et al. (2012). Expanding the frequency range of the TEM-t cell for the measurement of shielding materials up to 12 GHz. *International Symposium on Electromagnetic Compatibility-EMC EUROPE*. IEEE.

16 De Smedt, R. et al. (1995). Numerical characterization of a TEM-t cell as evaluation case of field solvers. *Proceedings of International Symposium on Electromagnetic Compatibility*. IEEE.

17 Celozzi, S., Araneo, R., and Lovat, G. (2008). *Electromagnetic Shielding*. Wiley.

18 Chung, Y.-C., Lee, J., Kwun, S.-T. et al. (2014). Comparison of SE evaluation methods for HEMP shelters. *The Journal of Korean Institute of Electromagnetic Engineering and Science* 25 (11): 1197–1200.

19 Sheet, M. S. (2005). High-Altitude Electromagnetic Pulse (HEMP) Protection for Ground-based C41 Facilities Performing Critical, Time-urgent missions Part, 1.

20 Svetanoff, D., Croisant, W., & Wehling, N. (2007) *IEEE Standard Method for Measuring the Effectiveness of Electromagnetic Shielding Enclosures. IEEE Std*, 299-2006, p. 299–2006.

21 Radasky, W.A. and Hoad, R. (2020). Recent developments in High Power EM (HPEM) standards with emphasis on High Altitude Electromagnetic Pulse (HEMP) and Intentional Electromagnetic Interference (IEMI). *IEEE Letters on Electromagnetic Compatibility Practice and Applications* 2 (3): 62–66.

22 Radasky, W.A. and Longoria, S.N. (2017). Recommended improvements for MIL-STD-188-125-1. *2017 XXXIInd General Assembly and Scientific Symposium of the International Union of Radio Science (URSI GASS)*. IEEE.

23 Seo, M., Chi, S., Kim, Y. et al. (2014). Electromagnetic wave shielding effectiveness measurement method of EMP protection facility. *The Journal of Korean Institute of Electromagnetic Engineering and Science* 25 (5): 548–558.

24 Yaw, K.C. (2012). Measurement of dielectric material properties, *Application Note. Rohde & Schwarz*, 1–35.

4

Graphene and Its Derivative for EMI Shielding

4.1 Introduction

Owing to its excellent electrical conductivity, electron mobility, optical transparency, and mechanical properties, graphene has found numerous applications in electromagnetic (EM) shielding and microwave absorbing materials. Graphene consists of a single atomic layer of carbon atoms packed into a two-dimensional (2D) honeycomb structure [1]. The carbon atoms in the graphene lattice are connected by sp^2-hybridized bonds, with the s, p_x, and p_y orbitals of the carbon atoms forming σ-bonds in the in-plane direction (between neighboring carbon atoms) and the p_z orbitals forming π-bonds in the out-of-plane direction, constituting the bonding (π) and antibonding (π^*) bands of graphene [2]. These π and π^* bands touch each other at the Fermi energy level at the K-point in the Brillouin zone, resulting in two conical points (K and K') per Brillouin zone where band crossing can occur. The π bands originating from the extra electrons available in the graphene lattice are freely available for electron conduction in three-dimensional (3D) space and are largely responsible for the outstanding electron conduction of graphene. These π bands also provide weak interactions between graphene layers and a substrate [2, 3].

The overlap point between the valence and conduction bands of graphene at the Dirac point results in this material having a zero bandgap. The electron band structure of graphene resembles that of metals, owing to the disappearance of the Fermi surface. It also resembles that of semiconductors, owing to the absence of a bandgap at room temperature [4]. These attributes make the charge carriers in graphene relativistic particles that behave as massless Dirac fermions. The electron–hole symmetry and the internal degree of freedom give rise to interesting electronic properties such as the quantum Hall effect, the ambipolar electric field effect, and the ballistic conduction of charge carriers [5].

In their pioneering work, Geim and coworkers realized an electric field effect and demonstrated that single-layer graphene acts as a field-effect transistor and ambipolar conductor with a conductivity of up to $10\,000\,cm^2\,V^{-1}\,s^{-1}$ [6]. Similarly, the integer quantum Hall effect originating from the Dirac nature of the bands of bilayer graphene was reported by Geim and coworkers [7]. Bandgap tuning was demonstrated for the first time by Ohta et al. through potassium doping to realize an n-type electronic switch [8]. Soon after, a number of studies reported the doping

Two-Dimensional Materials for Electromagnetic Shielding, First Edition.
Chong Min Koo, Pradeep Sambyal, Aamir Iqbal, Faisal Shahzad, and Junpyo Hong.

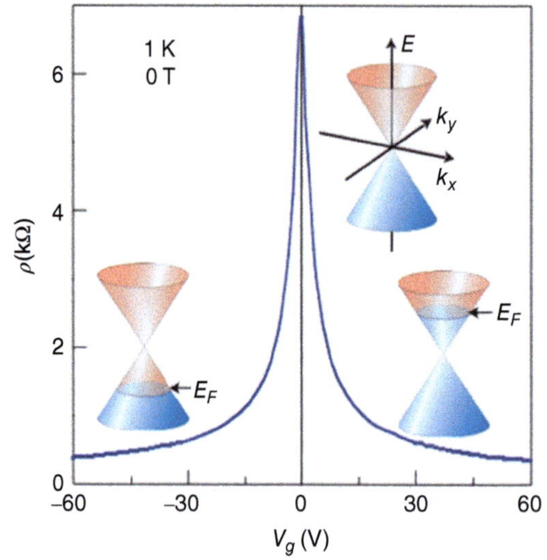

Figure 4.1 Ambipolar electric field effect in single-layer graphene. The rapid decrease in resistivity ρ on adding charge carriers indicates their high mobility (in this case, $\mu \approx 5000\,cm^2\,V^{-1}\,s^{-1}$ with no noticeable changes with increasing temperature up to 300 K). Source: Geim et al. [12]. Reproduced with permission of Nature Publishing Group.

of graphene using a variety of dopants to create p-type and n-type devices [9, 10]. Furthermore, the quantum Hall effect was demonstrated at room temperature in 2007 [11], which actively progressed graphene research toward device miniaturization. A major breakthrough in electronic transport in graphene occurred when an electron mobility approaching 200 000 $cm^2\,V^{-1}\,s^{-1}$ was revealed, which provided evidence for ballistic transport at liquid helium temperatures (Figure 4.1) [13].

In addition to possessing high electronic mobility, graphene also exhibits the highest known thermal conductivity to date. Balandin and coworkers measured the thermal conductivity of single-layer graphene using an indirect approach based on peak shifting in Raman spectra and extracted a thermal conductivity as high as 5300 $W\,m^{-1}\,K^{-1}$ [14]. The high electrical and thermal conductivity has given rise to thermal heater applications powered by the Joule heating phenomenon. Thus, graphene is an excellent material for thermal management applications. Similarly, graphene is the strongest material known to mankind with a Young's modulus of 1 TPa and an intrinsic strength of 130 GPa [15]. Graphene also exhibits 97.7% transparency, making it an ideal material for optoelectronic applications [16]. As the majority of graphene research utilizes the wet processing route, the surface of graphene is rich in tunable functional groups that can provide different properties. In particular, surface functionalization is ideal for producing stable dispersions in a variety of solvents, which is greatly desirable for applications in energy storage devices, electrochemical sensing, biomedicine, drug delivery, organic electronics, mechanical reinforcement additives, and fuel cells [17–19]. The surface functionalization of graphene is also useful for developing polymer composites for a variety of applications. Similarly, graphene possesses the largest surface area known for any material, which is highly useful for developing catalytic applications for the oxygen reduction reaction, hydrogen evolution reaction, oxygen evolution reaction, nitrogen reduction reaction, and others [20–22].

Since its discovery in 2004 by mechanical exfoliation, over a dozen different ways to produce graphene have emerged [23]. The progress in graphene synthesis is not limited to different production routes but also includes different source materials. Each method has some advantages and a few limitations. For example, mechanical exfoliation provides the highest quality defect-free graphene, but it is not a scalable process. Although chemical vapor deposition (CVD)-grown graphene films can be produced on a large scale, realizing roll-to-roll and batch-to-batch processes, the energy consumption and cost are considerable. The epitaxial growth of graphene on Si offers a high-quality product with a unified crystallographic orientation; however, the high cost of SiC wafers and further device integration pose great challenges. The effect of the synthesis process is evident from the quality of graphene products in the market, which contain approximately 60% sp^2 bonds along with several contaminants [24]. The high-impurity content coupled with the intrinsic defects in graphene leads to a deterioration of the properties expected for pure graphene. Furthermore, the uniformity and reproducibility of graphene are crucial for designing state-of-the-art miniaturized electronic applications [25]. Therefore, synthesis methods that are scalable and produce defect-free graphene materials in high yields are required for applications such as electromagnetic interference (EMI) shielding.

In recent years, there has been a phenomenal growth in the use of telecommunication devices that emit or receive electromagnetic waves (EMWs). The performance and durability of these telecommunication devices are strongly affected by stray electrical signals and the quantum of unwanted electromagnetic radiation surrounding electronic devices. These problems can be minimized or eliminated through the use of EMI shielding materials. When EMWs strike the surface of a material, they are either reflected, absorbed, or transmitted through the material. The extent of reflection or absorption depends on the material properties, such as electrical conductivity, permittivity, permeability, thickness, and frequency. The shielding properties can thus be tuned by changing any of these parameters.

Historically, many kinds of shielding materials have been used and explored, including metals, ferrites, conducting polymers, polymer composites, and carbon materials [26]. All these materials have certain advantages and disadvantages. Among current shielding materials, carbon-based materials, including carbon nanotubes (CNTs), carbon fibers, carbon black, and graphite, are the most notable. Electrical conductivity plays a crucial role in defining the ability of a material to shield EMWs. Graphene, owing to its ballistic transport properties, can reach electrical conductivities as high as $3000\,S\,cm^{-1}$ and has been explored in a variety of forms, including polymer composites, for different EMI shielding applications. Furthermore, excellent thermal conductivity coupled with superb electrical conductivity, solution processing ability, corrosion resistance, and the ability to be embedded in polymer matrices to realize conductive polymer composites make graphene a leading material for thermal heaters, heat-dissipating materials, thermal interface materials, and EMI shielding applications [27, 28].

Importantly, the unique 2D architecture of graphene is largely responsible for its utilization in forming polymer composites. 2D materials can form percolation networks at low filler contents, thereby reducing the amount of filler needed to obtain a polymer composite. Interestingly, 2D fillers such as graphene, transition metal carbides/nitrides/carbonitrides (MXenes), and transition metal dichalcogenides (TMDCs) all possess low densities that are only slightly higher than those of polymers, which makes them ideal for filler loading in polymer composites as compared to metallic fillers. This small difference in density also improves dispersion in polymer matrices. 2D materials have large surface area, which means they can achieve greater surface exposure toward incoming EMWs than zero-dimensional or one-dimensional (1D) fillers. Furthermore, the 2D nature of these materials provides numerous opportunities for surface functionalization to construct high-strength materials through strong interfacial bonds between the filler and the host polymer matrix. The dispersion ability of 2D materials in different solvents at varying concentrations could help in formulating conductive inks. Thin films made from 2D materials could be formed using layer-by-layer architectures to realize high-performance EMI shielding materials. Thus, 2D materials possess a number of advantages over other materials, particularly for EMI shielding applications.

This chapter focuses on the application of graphene in EMI shielding materials and microwave absorbing materials in different forms, such as graphene films, transparent graphene architectures, graphene–polymer composites, graphene hybrids with other materials, doped graphene, and graphene fillers with magnetic materials. The latest design strategies along with synthesis methods are discussed with an emphasis on improving the EMI shielding and microwave absorption properties of graphene.

4.2 Graphene for EMI Shielding

Graphene has become the most widely researched material for EMI shielding over the past decade, owing to its excellent electrical conductivity, solution processability, corrosion resistance, lightweight, and ability to be embedded in polymer matrices to form polymer composites. Graphene has been used for EMI shielding and microwave absorbing applications in a variety of forms including pristine thin films, sandwiched films, and fillers in polymers, aerogels, foams, and other architectures. This section provides a brief overview of each type of graphene structure along with a discussion of the corresponding shielding mechanisms.

4.2.1 CVD-Grown Graphene Films with Transparency

Graphene produced by mechanical exfoliation has outstanding physical properties; however, this approach is not viable for producing graphene in bulk quantities. Graphene produced via the CVD process is most similar to mechanically exfoliated graphene in terms of physical properties, mainly because of a low content of structural defects, which are commonly introduced by other methods such as chemical exfoliation. CVD-grown graphene can also be effectively transferred onto different substrates. The bulk-scale synthesis of CVD-grown graphene is difficult;

nevertheless, when transparency is an essential requirement, CVD-grown graphene films could dominate the commercial market owing to the high intrinsic transparency of monolayer graphene [29–31]. Conventional conductive thin films such as indium tin oxide (ITO) or fluorine-doped tin oxide (FTO) only exhibit moderate shielding performance at the desired transparencies [32]. In contrast, graphene exhibits good optical transparency while maintaining a reasonable shielding level.

Hong et al. reported the EMI shielding effectiveness (SE) of monolayer graphene produced by CVD [33]. CVD-grown graphene was transferred to a transparent polyethylene terephthalate (PET) film and used for subsequent EMI shielding measurements. As a reference, the pristine PET film showed a negligible EMI SE. Figure 4.2a shows the EMI SE, absorbance, and reflectance of monolayer graphene. The results indicated that a graphene monolayer can shield approximately 40% of the incident EMWs, with absorption being the dominant phenomenon. To investigate the influence of defects, the authors synthesized defective graphene with a large number of defects and tested the EMI SE. Owing to a very high sheet resistance ($R_s \approx 2.3\,\text{k}\Omega\,\text{sq}^{-1}$; Figure 4.2d) as compared to the graphene monolayer control ($R_s \sim 635\,\Omega\,\text{sq}^{-1}$; Figure 4.2c), the defective graphene exhibited no EMI shielding capability (Figure 4.2b). The high-intensity D-band peak in the Raman spectrum (Figure 4.2d) was a clear indication of the large number of defects in the defective graphene as compared to the reference sample (Figure 4.2c). For comparison, the authors also determined the shielding performance of thin metal films with three different thicknesses (10, 15, and 20 nm) made by evaporating Au, which had EMI SE values of 9.11, 13.8, and 19.51 dB, respectively (Figure 4.2e). The results indicated that the EMI SE of three-layer graphene (thickness ~1 nm) is similar to that of a 10 nm Au film. The slope of the curve indicated that the EMI SE values per unit thickness were 7.73 and $1.014\,\text{dB}\,\text{nm}^{-1}$ for graphene and Au films, respectively. Thus, the SE of graphene was almost seven times that of Au films with similar thicknesses. These experimental results were compared with theoretical calculations using electrical conductivity data. The calculated EMI SE was in excellent agreement with the measured EMI SE for both graphene and the Au films, as shown in Figure 4.2f. It is important to note that defect-free monolayer graphene, with an intrinsic conductivity of $\sim 10^8\,\text{S}\,\text{m}^{-1}$, can theoretically exhibit an EMI SE of ~16.5 dB [34]. However, graphene synthesis via chemical routes induces defects that significantly deteriorate the electrical conductivity by several orders of magnitude. Thus, the high EMI SE expected for ideal monolayer graphene is yet to be realized.

The graphene produced by common synthesis routes exhibits EMI SE values in the range of ~2–3 dB [33]. To enhance the shielding performance while retaining optical transmittance, graphene has been hybridized with other metallic materials, such as Al, Ni, and Ag [35, 36]. In one such work, Ma et al. fabricated conductive graphene hybrid films consisting of a metallic network for EMI shielding [37]. The graphene film was made on a Cu foil at 1000 °C, followed by room temperature etching of Cu and coating with poly(methyl methacrylate) (PMMA) (Figure 4.3a). Separately, a 400–450 nm thick Al layer was sputter coated onto a quartz glass substrate and covered with a photoresist. The photoresist was subsequently removed

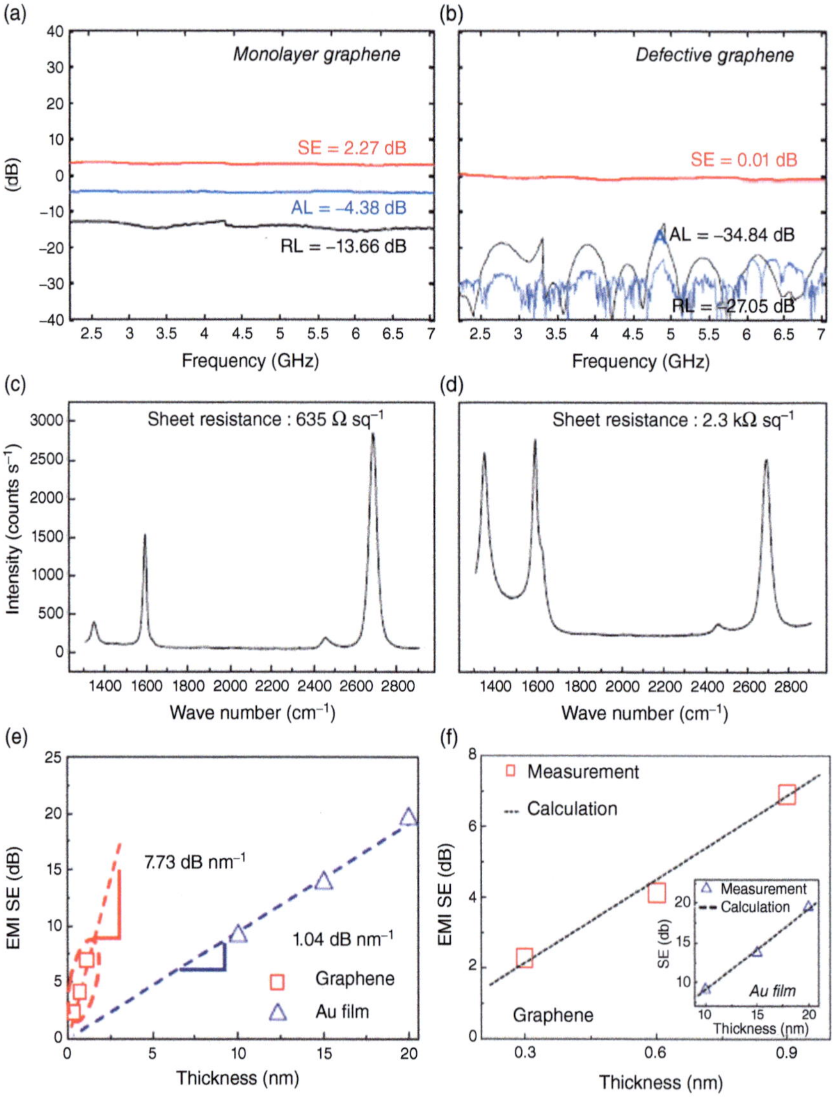

Figure 4.2 EMI SE, absorbance loss (AL), and reflection loss (RL) of (a) monolayer graphene and (b) defective graphene. Raman spectra of (c) monolayer graphene and (d) defective graphene. (e) Comparison of the EMI SE of graphene and Au thin films. (f) Comparison of the measured EMI SE of graphene and Au films with the EMI SE calculated using the plane-wave theory at a single frequency of 2 GHz. Source: [33]. Reproduced with permission of IOP Publishing.

from the Al metal mesh, which was stamped on the PMMA/graphene sample. Subsequently, PMMA was also removed, leaving behind a hybrid graphene/Al metal mesh structure with a period of 160 or 320 μm (160 and 320 M, respectively).

Figure 4.3b,c shows photographs of the G/320 and G/160 M hybrid films, which are further illustrated in Figure 4.3d,e. Figure 4.3f shows the normalized

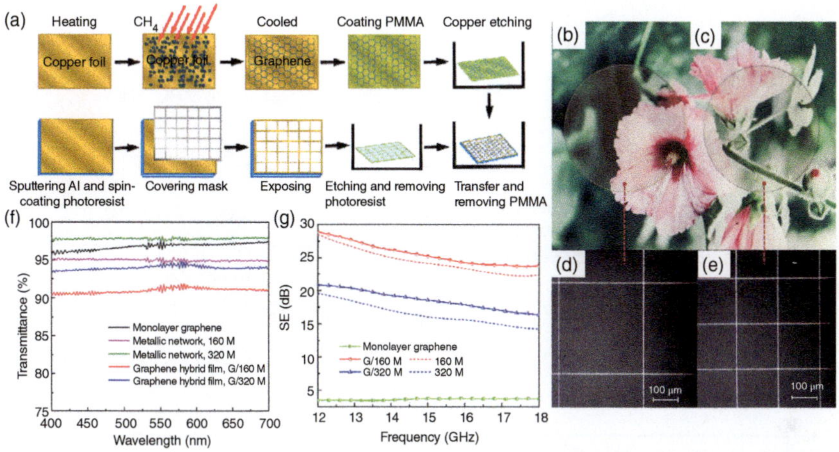

Figure 4.3 (a) Synthesis process for graphene hybrid film. (b, c) Qualitative demonstration of the transparency of G/320 and G/160 M laid over a photograph. (d, e) Micrographs of G/320 and G/160 M, respectively. (f) Normalized optical transmittance of monolayer graphene, metallic networks, and graphene hybrid films. (g) EMI SE of monolayer graphene, metal networks, and graphene hybrid films at 12–18 GHz. Source: Ma et al. [37]. Reproduced with permission of the American Chemical Society.

visible transmittance of the monolayer graphene, metallic network, and graphene hybrid films. The monolayer graphene film exhibited a transparency of ~97% at 400–700 nm, whereas the 160 and 320 M samples showed transparencies of 95% and 97.8%, respectively. The slightly lower transparency of pristine graphene (PG) was due to the intact PMMA layer, which could not be removed due to handling issues. However, because there was no PMMA layer in the case of the metal mesh/graphene structures, outstanding optical transparency was achieved. The EMI shielding performance of the monolayer graphene and hybrid structures was evaluated, as presented in Figure 4.3g. The EMI SE of the graphene hybrid film was higher than that of PG or the metal mesh, thus substantiating the combined role of graphene and the metal mesh in realizing good shielding materials. The graphene hybrid film yielded an EMI SE of ~24 dB with a visible transmittance of 91% or an EMI SE of ~16 dB with a visible transmittance of 94% at 12–18 GHz.

Kim et al. fabricated reduced graphene oxide (RGO) sheets interleaved between polyetherimide (PEI) films by electrophoretic deposition (EPD) [38]. The RGO layer was deposited on a stainless steel (SS) plate by anodic EPD, followed by q-PEI deposition by cathodic EPD to construct an RGO/PEI architecture. The RGO/PEI film was removed from the SS substrate with the help of scotch tape. Owing to the very small thickness of the RGO sheets, the RGO/PEI film was transparent and flexible. Using individual RGO/PEI films, a multilayer structure was fabricated, as shown in Figure 4.4a. Two RGO/PEI films were thermally compressed with one adhesive PEI film inserted between them to construct a PEI/RGO/PEI/RGO/PEI structure. Similarly, a three-layer RGO/PEI film was constructed by the addition of another RGO/PEI film and adhesive layer.

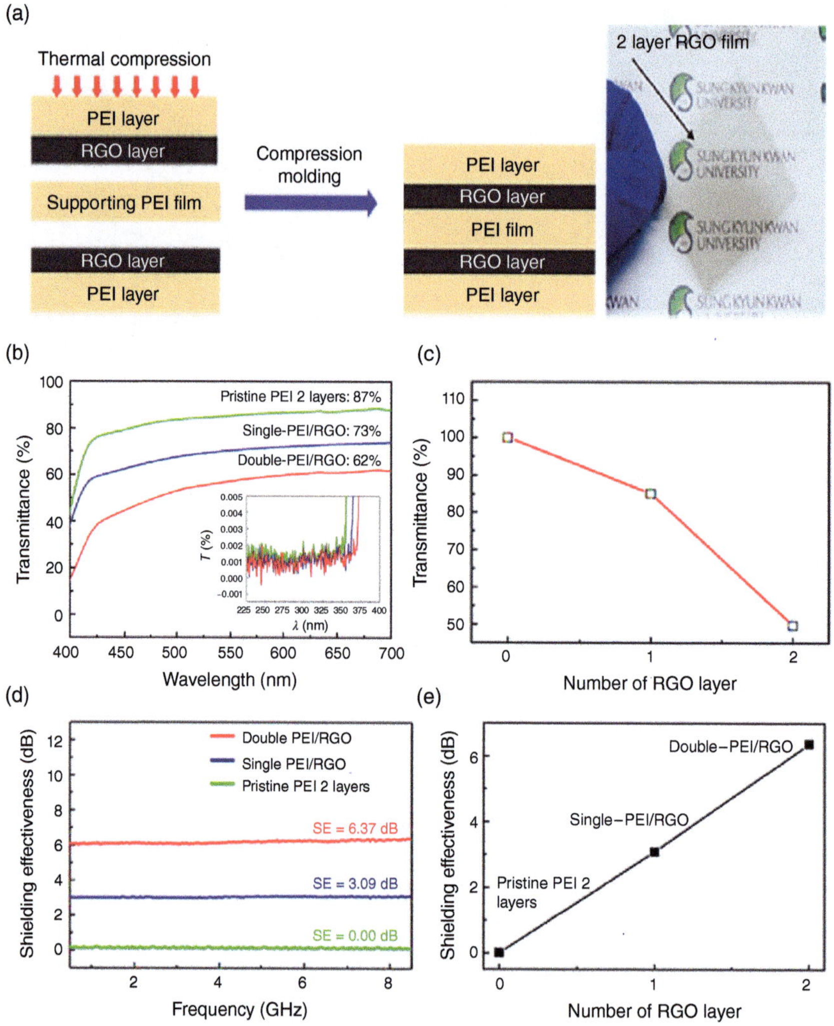

Figure 4.4 (a) Schematic of double-PEI/RGO film fabrication with a digital image of the resulting film. (b) Light transmittance of two layers of pristine PEI (green line), a single-PEI/RGO film (blue line), and a double-PEI/RGO film (red line). (c) Optical transmittance vs. number of RGO layers. (d, e) Comparison of the SE values of films with different numbers of RGO layers. Source: Kim et al. [38]. Reproduced with permission of the American Chemical Society.

The light transmittances of pristine PEI, PEI/RGO/PEI (single RGO layer), and PEI/RGO/PEI/RGO/PEI (double RGO layer) are shown in Figure 4.4b. Pristine PEI exhibited a transmittance of 87%, which gradually decreased to 62% after assembling the two RGO layers (Figure 4.4c). The electrical conductivity of the double PEI/RGO film was $1250\,S\,m^{-1}$. Furthermore, an increase in EMI shielding was observed with the number of RGO layers. The single PEI/RGO film exhibited an EMI SE of 3.09 dB, which was improved to 6.37 dB upon increasing the number of PEI/RGO films (Figure 4.4d,e). Shielding by absorption was found to be the dominant shielding mechanism, with 92% and 96% of the SE values (3.09 and 6.37 dB) attributed to the absorption contribution. The EMI SE increased almost linearly with the increase in the number of RGO layers. This approach could be used to assemble multiple layers of RGO to constitute architectures with a desirable EMI shielding ability and optical transparency.

Zhang et al. utilized a PMMA-assisted transfer method to fabricate transparent graphene films, as shown in Figure 4.5a [39]. The CVD-grown graphene was successfully transferred onto a transparent PET film and a glass substrate by the PMMA-assisted method to obtained one to three graphene layers, denoted as PET/graphene-1-PMMA, PET/graphene-2-PMMA, and PET/graphene-3-PMMA, respectively. The authors used a doping strategy to further improve the conducting properties. Specifically, HNO_3 was used as a p-type dopant, as an electron transferred from graphene to HNO_3 leaves a hole in the graphene lattice. The carrier concentration of graphene increased with a shift in the Fermi level, thus improving the electrical conductivity. The EMI shielding with and without doping revealed an improvement of almost 30–60% over the X-band range. The transferred graphene film also exhibited a high transmittance of up to 91%, which was slightly lower than that of an ideal graphene sheet (Figure 4.5b). Figure 4.5c shows the EMI SE of all the samples. The negligible shielding properties of the PET or PET/PMMA films were improved by the introduction of a graphene layer. The three-layer structure showed a significant EMI SE of ~10 dB at 9.5 GHz with an optical transmittance of over 80%. After the stacking of each additional graphene layer, the EMI SE increased by almost 2 dB, which suggests that shielding can be tuned for a given transmittance requirement. In this structure, the impedance mismatch between PMMA and graphene, multiple reflections at the interlayers, and the intrinsic high conductivity, which was further improved by doping, played significant roles in enhancing the EMI SE.

In another work, Lu et al. utilized a similar PMMA-assisted transfer method with CVD-grown graphene to construct multilayer graphene/PET stacked structures, as illustrated in Figure 4.5d [40]. The fabricated films were subjected to EMI shielding measurements, as shown in Figure 4.5e. The pristine PET film exhibited a negligible EMI SE. The average EMI SE of the eight-layer graphene/PET sample was 19.14 dB, which can block nearly 98.8% of incident EMWs. Obviously, the EMI SE increased as the number of graphene layers increased, whereas the optical transmittance decreased gradually. The six-layer graphene/PET sample (graphene thickness of ~3 nm) achieved an EMI SE of 14.73 dB and a visible transmittance of 84.7%. Similarly, the eight-layer graphene/PET sample exhibited a visible transmittance of 80.5%. In all the graphene structures, shielding by absorption was the dominant

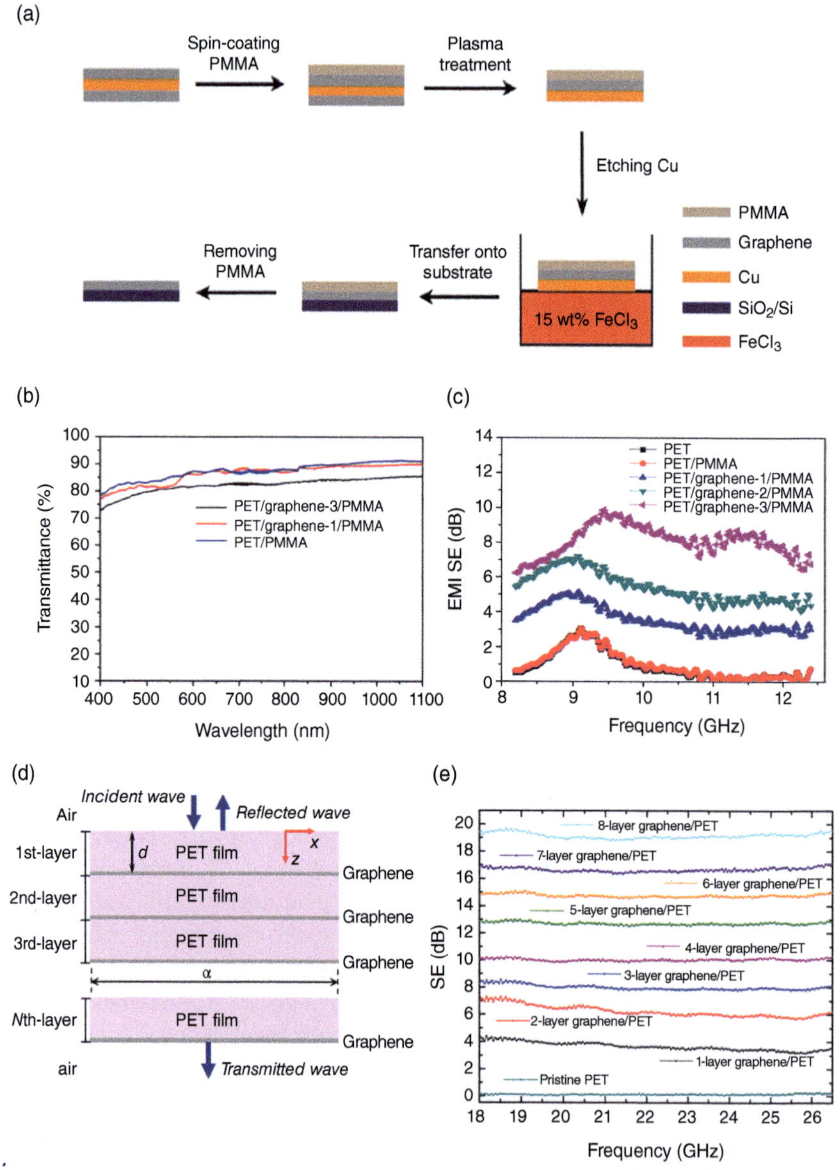

Figure 4.5 (a) Schematic diagram of the PMMA-assisted transfer method, (b) transmittance of PET/PMMA and PET/graphene-*N*/PMMA ($N = 1, 3$), and (c) EMI SE of all the samples. Source: [39]. Reproduced with permission of the American Chemical Society. (d) Description of wave dispersion in the multilayer graphene/PET structure and (e) EMI SE of the pristine PET film and multilayer graphene/PET samples measured at 18–26 GHz. Source: [40]. Reproduced with permission of the Royal Society of Chemistry.

mechanism. The extra interfaces between the graphene layers, together with the overall increase in the graphene thickness, led to high EMI SE values (Figure 4.5d).

4.2.2 Graphene Laminate Films

Despite significant progress, the use of CVD-grown graphene has not matured, mainly because of the technological challenges associated with large-area production as well as the high processing cost. For EMI shielding, pristine CVD-grown graphene cannot compete with other materials because of the limitations on making thicker graphene films, which are essential for enhanced shielding properties. Single-layer graphene does not provide sufficient shielding capability, whereas the production of thicker graphene films via CVD is not commercially feasible. Moreover, CVD-grown graphene cannot be easily embedded in polymer matrices to form polymer composites. Owing to these limitations, the majority of graphene research has focused on chemically derived graphene. Different strategies have been used, such as vacuum filtration [41], direct evaporation [42, 43], spray coating [44, 45], spin coating [46], and roll-to-roll production [47], to construct thin films with varying thicknesses for shielding purposes. Each of the aforementioned processes has advantages and limitations. For example, vacuum filtration is a viable approach for fabricating films with good mechanical and electrical properties, and films with variable thicknesses can be obtained to control the shielding properties on demand. However, this process is not scalable and generally requires a long time to construct a film. The direct evaporation or self-assembly approach is a useful low-temperature process; however, handling these films is a crucial issue in large-scale manufacturing. Spin coating is also an interesting approach for producing films with variable thicknesses; however, to use this method, graphene must be stably dispersed in the liquid phase. Because PG is hydrophobic, it must be functionalized to achieve stable dispersions, which compromises the electrical conductivity. Recently, the roll-to-roll production of graphene films has been realized for large-scale manufacturing and subsequently use in shielding technology. However, post-film-making steps, such as the chemical or thermal reduction of graphene oxide (GO), remain challenging. The need for special conditions for different film-making technologies, such as a vacuum/inert atmosphere, electrodes, centrifuges, and high temperatures, limits the scalability of these graphene films.

In this section, we highlight the major advances in the synthesis and development of graphene films for EMI shielding purposes. For example, Zheng and coworkers reported EMI shielding of a ~8.4 μm thick graphitized graphene film prepared via the direct evaporation of a GO solution at mild temperatures, followed by graphitization at higher temperatures [42]. Figure 4.6a shows the schematic of the GO film fabrication process, in which a GO suspension was poured into a Teflon dish and allowed to evaporate at low temperatures (50–60 °C) until dry (6–10 hours). Subsequently, a freestanding, dark-brown GO film (Figure 4.6b) was obtained with a controllable thickness, which was subjected to thermal annealing at 2000 °C to reduce and graphitize the GO film (denoted as GF-2000). The film appearance changed from dark brown to shiny metallic luster (Figure 4.6c), which

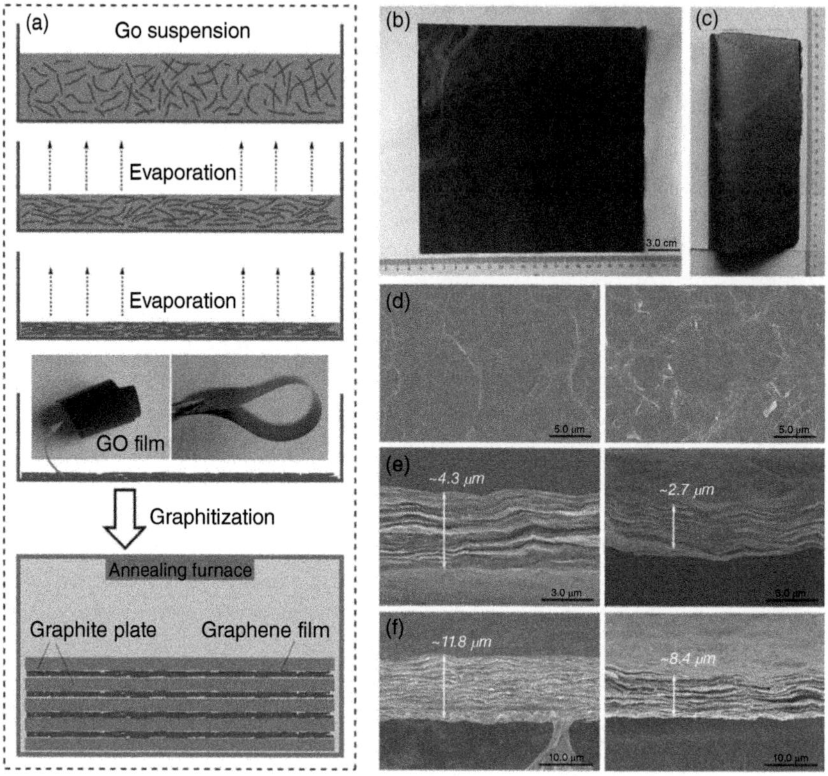

Figure 4.6 (a) Schematic representation of the self-assembly process for GO film preparation by the evaporation of a GO solution. Photographs of (b) a freestanding GO film and (c) a folded GF-2000 film. (d) Scanning electron microscopy (SEM) images showing the surface morphology of the GO film (left) and GF-2000 (right). (e, f) SEM images showing the layer-by-layer nanostructure in the cross sections of the GO film and GF-2000 (with different thicknesses). Source: Shen et al. [42]. Reproduced with permission of Wiley-VCH.

was an indication of good conducting properties. The GO film was smooth with thin ripples (Figure 4.6d) and the fractured edges revealed a layer-by-layer morphology in both the GO film (Figure 4.6e) and GF-2000 (Figure 4.6f).

After the graphitization treatment, the GF-2000 film exhibited an excellent electrical conductivity of 1000 S cm^{-1}, which should provide outstanding shielding properties. Consequently, the GF-2000 film was attached to a polyolefin elastomer (POE) using double-sided tape (Figure 4.7a), and EMI shielding measurements were conducted in the X-band (Figure 4.7b,c). As the GO film is insulating, it exhibited a very low EMI SE in the X-band range. On the contrary, the GF-2000 film showed an outstanding EMI SE of >10 dB at a very small thickness of 2.7 μm. The EMI SE was further improved by increasing the film thickness to 8.4 μm with an EMI SE of >20 dB at a frequency of 8.5 GHz. The EMI SE mechanisms were elucidated with the help of scattering parameters, which showed that both SE_A and SE_R contributed to the overall shielding; however, the absorption contribution was dominant in all the tested films. The authors also tested the thermal conductivity

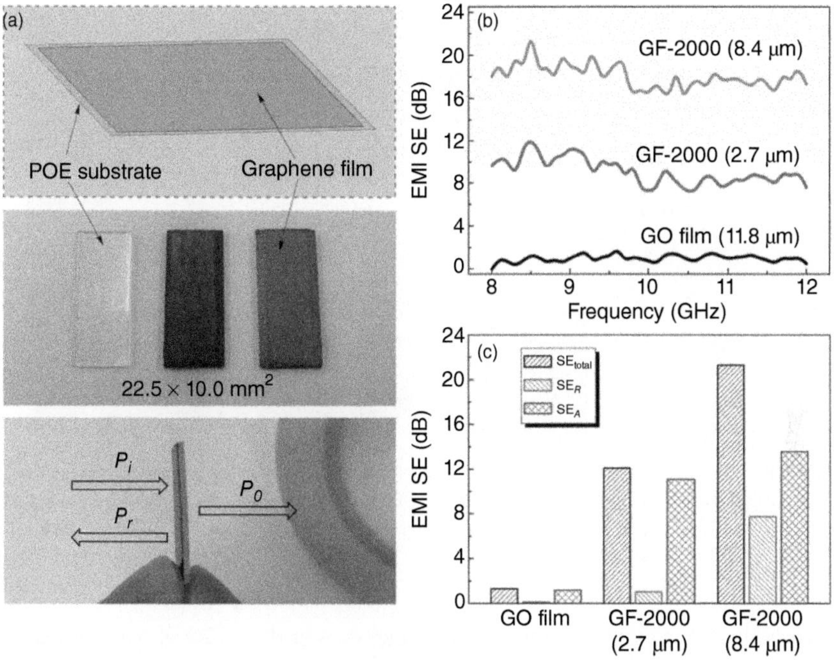

Figure 4.7 (a) Schematic representation of the POE/GF-2000 film with dimension and shielding mechanisms. (b) EMI SE of the GO film and GF-2000. (c) SE_T, SE_R, and SE_A of the GO film and GF-2000 at 8.5 GHz. Source: Shen et al. [42]. Reproduced with permission of Wiley-VCH.

of the graphitized film and a high thermal conductivity of ~1100 W m^{-1} K^{-1} was observed, which far exceeded that of pure copper (~400 W m^{-1} K^{-1}). This high thermal conductivity was attributed to reduced thermal contact resistance and phonon-boundary scattering resulting from the more compact packing of larger aligned graphene sheets with few defects in the GF-2000 film. The graphitized film also exhibited good mechanical flexibility and structural integrity during bending.

Kumar et al. prepared freestanding GO and RGO films via a vacuum filtration method. They studied the effect of the lateral sheet size of graphene on electrical conductivity, thermal conductivity, and EMI shielding [41]. Initially, a large graphite flake (>80 μm) was used, and after the chemical exfoliation and centrifugation steps, large sheets were collected, denoted as large graphene oxide (LGO) (Figure 4.8a). Furthermore, the graphene sheets were broken into smaller pieces, denoted as small graphene oxide (SGO), by an additional sonication step, as depicted in Figure 4.8b. The LGO films, which had a typical brown color (Figure 4.8c), were reduced with hydroiodic (HI) acid to obtain rLGO films with a metallic silver color, as shown in Figure 4.8d, while retaining mechanical flexibility (Figure 4.8e). The reduction treatment resulted in a decrease in film thickness (~7.5 μm for rLGO and ~10 μm for LGO), as shown in Figure 4.8f,g. The decreased film thickness was ascribed to the removal of adsorbed water and other oxygenated and hydroxyl groups, which

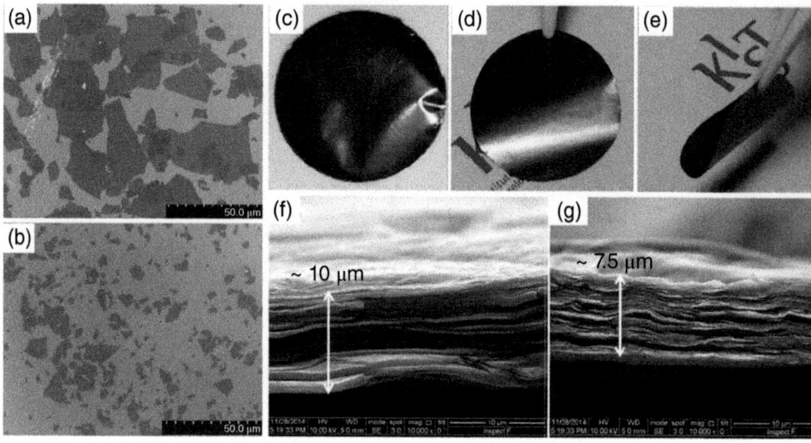

Figure 4.8 SEM images of (a) LGO and (b) SGO. Photographs of (c) the freestanding LGO film and (d, e) the shiny metallic and flexible rLGO film. (f, g) Cross-sectional SEM images of the as-prepared LGO and rLGO films, respectively. Source: Kumar et al. [41]. Reproduced with permission of Elsevier.

decreased both the overall weight and the thickness of the rLGO film as compared to the pristine LGO film.

Interestingly, the rLGO film exhibited better mechanical properties than the rSGO film, as shown in Figure 4.9a. The rLGO film had a Young's modulus of 6.3 GPa and a tensile strength of 77.7 MPa, which were larger than those of the rSGO film (Young's modulus ~4.6 GPa; tensile strength ~65 MPa). Similarly, the electrical conductivity and thermal conductivity of the rLGO film were better than those of the rSGO film. The rSGO film exhibited an electrical conductivity of ~152 $S\,cm^{-1}$ and a thermal conductivity of ~900 $W\,m^{-1}\,K^{-1}$ as compared to ~243 $S\,cm^{-1}$ and ~1390 $W\,m^{-1}\,K^{-1}$ for the rLGO film (Figure 4.9b). The better electrical and thermal properties of the rLGO film were ascribed to smaller d-spacing values, which resulted in more compact and aligned GO layers and a decreased intersheet contact resistance owing to the large lateral size of the graphene sheets. Similarly, the thermal conductivity value increased with the increase in graphene flake size because acoustic phonons with longer wavelengths were available for heat transfer. The rLGO film also exhibited an excellent EMI SE of ~20 dB at a thickness of ~15 μm, which was higher than the EMI SE of 17 dB obtained for an rSGO film with a similar thickness (Figure 4.9c). Furthermore, the effect of thickness was also demonstrated, with increases in EMI SE observed with increasing film thickness for both the rLGO and rSGO films (Figure 4.9d). The shielding mechanisms were also elucidated by determining the SE_R and SE_A contributions. The absorption contribution was dominant for both types of films (Figure 4.9e). The flexible nature of the RGO films coupled with their high thermal conductivities and EMI shielding properties provide an ideal platform for state-of-the-art EMI shielding applications.

In another study, Wei et al. developed a scalable scanning centrifugal casting (SCC) method to develop PG films and their polymer composites [48]. In this method, a PG dispersion was fed via step-by-step scanning injection onto a release film substrate

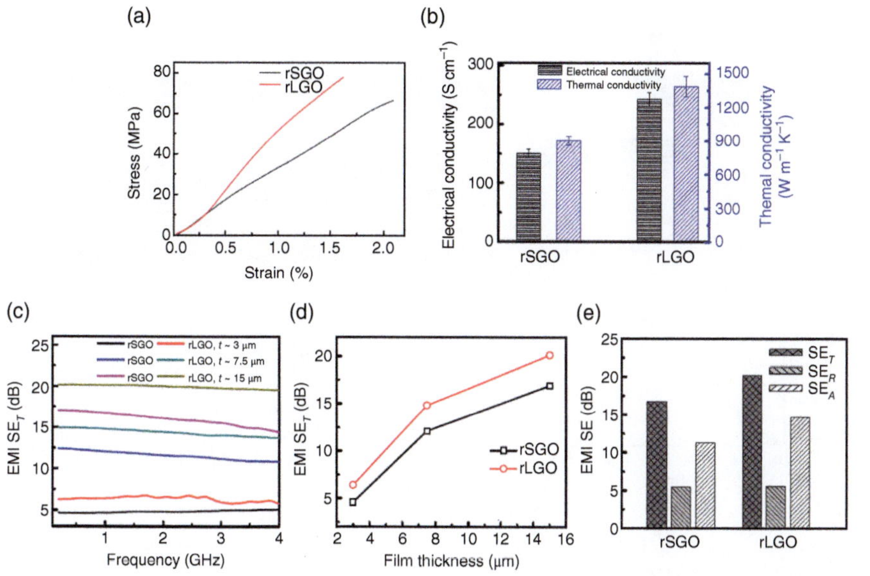

Figure 4.9 (a) Stress–strain curves, (b) electrical and thermal conductivities, and (c) EMI SE of rSGO and rLGO thin films. (d) Thickness-dependent EMI SE_T at 1 GHz and (e) EMI SE_T, SE_R, and SE_A of 15 μm thick samples at 1 GHz. Source: Kumar et al. [41]. Reproduced with permission of Elsevier.

attached to the inner wall of a rotating hollow tube (RHT), as shown in Figure 4.10a. After stopping the injection of liquid, the liquid film on the substrate was dried, and the thickness of the layer could be increased by injecting and drying more liquid. Thus, by repeating the SCC process, highly aligned and compact laminated PG films with tunable thicknesses were obtained, as depicted in Figure 4.10b–e. The alignment and compaction rate of the PG film could be improved by increasing the rotation rate of the RHT. In terms of its advantages, the SCC method is scalable and dust free; it produces no raw material waste, and it can be used to fabricate films with uniform thicknesses and smoother surfaces. More importantly, this method can be used to fabricate thin films of other 2D materials. The PG films exhibited high mechanical strength (~145 MPa), electrical conductivity (~1340 S cm^{-1}), and thermal conductivity (~190 W m^{-1} K^{-1}), which were impressive when compared to the properties of graphene produced by other methods (Figure 4.10f,g) [42]. The excellent tensile strength of the PG film was ascribed to the high intrinsic strength of crystalline PG, the alignment, and compaction-induced strong van der Waals forces and π–π interactions between the PG sheets.

The well-aligned PG sheets exhibited outstanding EMI shielding properties as a function of thickness, as shown in Figure 4.10h–j. The rotation rate had a significant effect on the EMI shielding and electrical conductivity of the PG films. A high rotation rate resulted in more compact and aligned films, which exhibited better shielding properties. Figure 4.10i exhibits the EMI SE for PG films with different thicknesses, as obtained by repeating the SCC process. Clearly, increasing the film thickness significantly improved the EMI SE. A very high value of 38.1 dB was obtained at a very small thickness of ~4 μm, which exceeds commercial

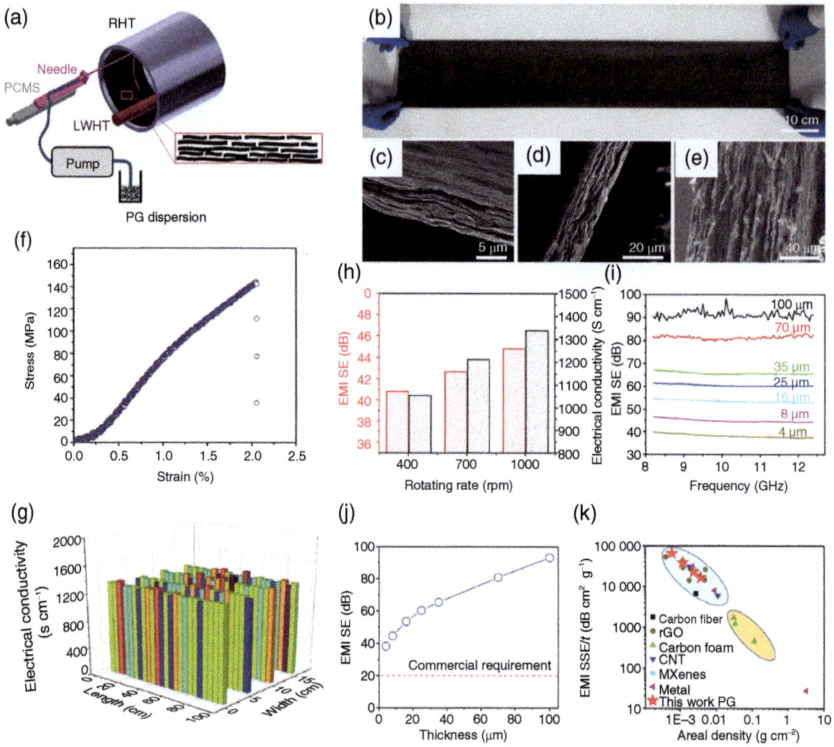

Figure 4.10 (a) Schematic illustration of the synthesis process of PG films by SCC. (b) A meter-scale freestanding PG film with a thickness of ~20 μm. (c–e) Cross-sectional SEM images of PG films with different thicknesses. (f) Stress–strain curve of a PG film. (g) Electrical conductivity of the large-area PG film. (h) EMI SE of PG films at different rotating rates (400, 700, and 1000 rpm). (i) EMI SE of PG films with different thicknesses. (j) EMI SE of PG films as a function of thickness at 10 GHz. (k) Comparison of EMI SSE/*t* of PG films with those of other materials. Source: Wei et al. [48]. Reproduced with permission of Wiley-VCH.

requirements and the performance of many other materials with comparable thicknesses (Figure 4.10j) [49–51]. Ideally, EMI shielding materials are lightweight or have low densities with good shielding performance at small thicknesses, which can be evaluated using the specific shielding effectiveness (SSE/*t*). The PG film showed an extremely high SSE/*t* value of 37 332 dB cm^2 g^{-1} at a very small thickness of ~4 μm and a low aerial density of 0.001 19 g cm^{-2}, which is higher than those of commercially used Al foil (30 555 dB cm^2 g^{-1} at ~8 μm) and Cu foil (7812 dB cm^2 g^{-1} at ~10 μm) (Figure 4.10k). The EMI SE performance originated from both the reflection and absorption contributions, where the multiple interfaces between PG sheets played a significant role in the attenuation of EMWs.

In another work, Cao and coworkers developed freestanding conductive and magnetic graphene-based paper by utilizing graphene nanosheets (GNs) to directly grow Fe_3O_4 magnetic nanoparticles [52]. The magnetic nanoparticles were grown over the GNs by a simple hydrothermal treatment, followed by filtration to construct a freestanding Fe_3O_4/GN conductive paper. Interestingly, the Fe_3O_4/GN conductive paper could be peeled off the filtration membrane, similar to the pristine GN paper.

Different samples with varying amounts of Fe_3O_4 nanoparticles on the GNs were prepared and designated as Fe_3O_4/GN-1 paper (16 wt% Fe_3O_4), Fe_3O_4/GN-1 paper (28 wt% Fe_3O_4), Fe_3O_4/GN-1 paper (37 wt% Fe_3O_4), and Fe_3O_4/GN-1 paper (50 wt% Fe_3O_4). The electrical conductivity was found to decrease with the increase in magnetic filler content, from 300 S cm^{-1} for pristine GNs to 120, 66, 57, and 50 S cm^{-1} for the Fe_3O_4/GN samples. However, the magnetic filler induced good magnetic properties, and the Fe_3O_4/GN-1 paper (50 wt% Fe_3O_4) showed a magnetization of 47 emu g^{-1}. The magnetic and conductive graphene hybrid paper benefitted from the electrical conductivity of the GNs and the magnetic properties of the Fe_3O_4 nanoparticles, resulting in a material with outstanding EMI SE properties. Generally, for ideal conductors, EMWs are reflected from the surface with only a small portion of the EMWs penetrating the material; however, for Fe_3O_4/GN paper, the EMWs penetrated the surface and were subsequently absorbed. In this case, the refractive index of the material was determined by both the relative permittivity and the relative permeability. Therefore, owing to competition between the electrical and magnetic parts for EMW mitigation, the Fe_3O_4/GN-1 paper (37 wt% Fe_3O_4) exhibited better SE_R properties, as shown in Figure 4.11a,b.

The absorption contribution was largely influenced by the addition of Fe_3O_4 nanoparticles. The EMWs were consumed by both electrical and magnetic losses as a result of the leakage current in the graphene-based conductive network and the loss caused by the magnetic dipoles in the Fe_3O_4 particles. Consequently, the increase in Fe_3O_4 content monotonically improved SE_A, as shown in Figure 4.11c,d. The highest SE_A value was recorded for Fe_3O_4/GN-1 paper (50 wt% Fe_3O_4) in the range of 13–17 dB. The measurements of the total EMI SE (Figure 4.11e,f) showed that the optimal EMI SE of 21–24 dB was obtained at a small thickness of 0.25 mm. In the Fe_3O_4/GN paper, the introduced Fe_3O_4 NPs were 10- to 100-fold larger than the thickness of a single GN sheet, which greatly affected the interfaces and interactions in the Fe_3O_4/GN hybrid paper (Figure 4.11g). It is important to note that the EMI SE values of PG papers were better than or comparable to those of the magnetic and conductive graphene paper; however, the shielding contributions from SE_R and SE_A were different in each case. Furthermore, the Fe_3O_4/GN hybrid paper (50 wt% Fe_3O_4) had a lower density (~0.78 g cm^{-3}) than the neat GN paper (1.6 g cm^{-3}) owing to the introduction of Fe_3O_4 nanoparticles between the GN sheets. The lightweight of the Fe_3O_4/GN hybrid paper resulted in a density similar to that of commonly used polymeric foams at very small thicknesses. Thus, the introduction of Fe_3O_4 nanoparticles into GN sheets increased the shielding performance, as depicted in Figure 4.11h, by affecting surface reflections, leakage current, and magnetic loss as well as through attenuation induced by the anisotropic character.

4.2.3 Graphene–Polymer Composites

PG laminates or films can provide good shielding properties; however, the material cost, environmental susceptibility, and mechanical properties are not always sufficient for direct utilization in the desired applications. In particular, graphene films with thicknesses of several micrometers could be very expensive, and their surfaces

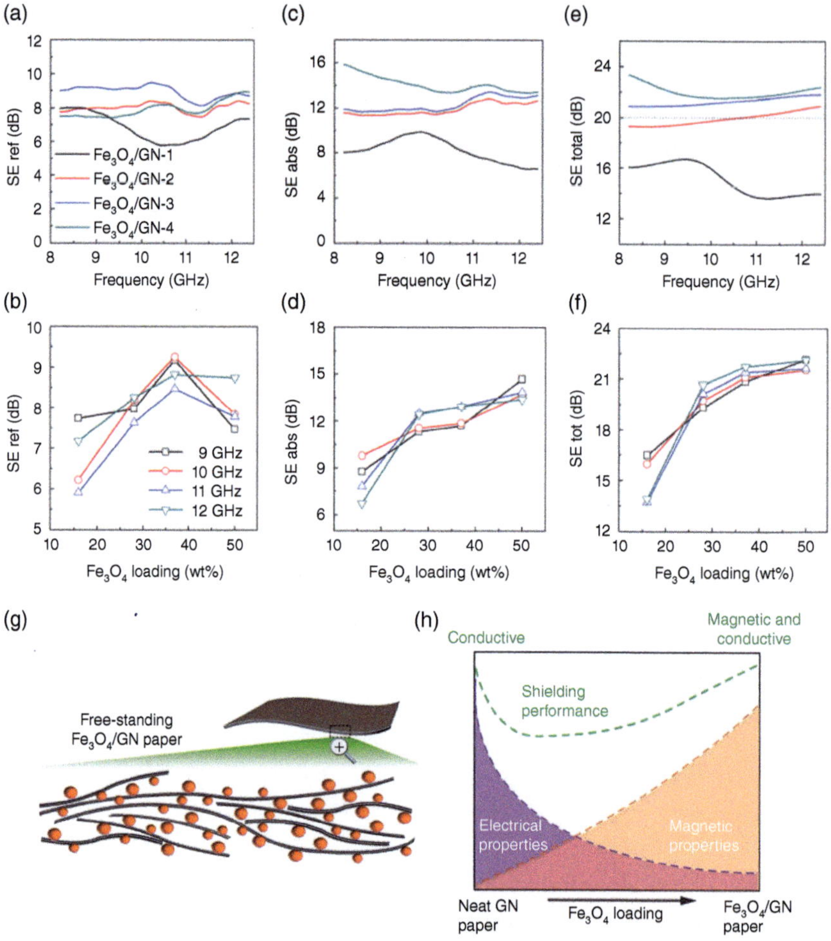

Figure 4.11 Dependence of (a) SE_R, (c) SE_A, and (e) SE_T of the Fe_3O_4/GN paper on frequency. Dependence of (b) SE_R, (d) SE_A, and (f) SE_T of the Fe_3O_4/GN paper on Fe_3O_4 loading. (g) Possible interactions between GN layers and Fe_3O_4 nanoparticles in the Fe_3O_4/GN paper. (h) Shielding performance schematic. Source: Song et al. [52]. Reproduced with permission of the Royal Society of Chemistry.

would be exposed to harsh environments in real applications. A better alternative is to develop polymer composites that utilize a very small amount of conductive filler, thus significantly lowering the cost and providing protection from environmental effects as well as reasonable mechanical strength [51, 53, 54].

Chen and coworkers reported the first graphene–polymer composite for EMI shielding [55]. They developed graphene/epoxy composites using an in situ process. An epoxy hardener was added to a suspension of partially reduced GO, which was then sonicated for several hours. The as-mixed product was poured into molds and cut into the desired shape before annealing at 250 °C for two hours under N_2 to fully reduce GO. Polymer composites with varying contents of RGO filler were fabricated to investigate the electrical conductivity and EMI SE properties.

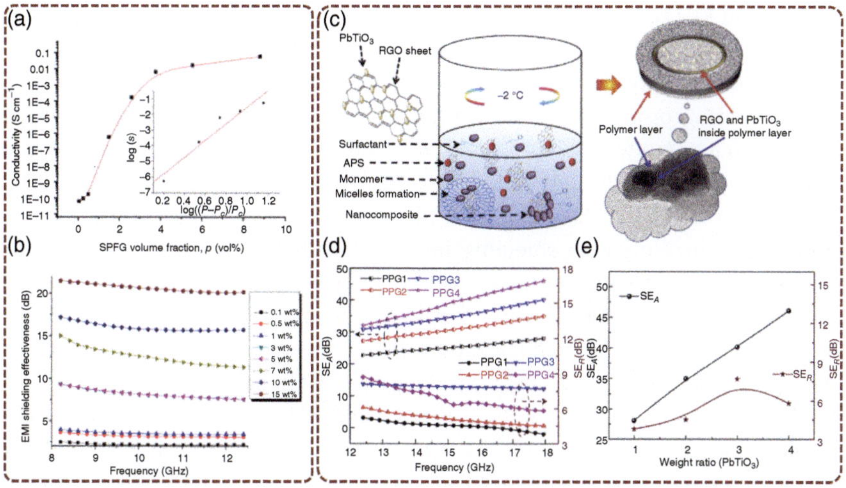

Figure 4.12 (a) Electrical conductivity of the graphene/epoxy composites as a function of SPFG volume fraction. (b) EMI SE of graphene/epoxy composites at various SPFG loadings. Source: Liang et al. [55]. Reproduced with permission of Elsevier. (c) Schematic of the synthesis of PPbG composites. (d, e) SE_A and SE_R of PPbG composites at different $PbTiO_3$ concentrations in the range of 12–18 GHz and at 18 GHz, respectively. Source: Dalal et al. [56]. Reproduced with permission of Elsevier.

Figure 4.12a shows the electrical conductivity as a function of the volume fraction of solution-processable functionalized graphene (SPFG). The electrical conductivity variation agreed well with the percolation behavior, with a percolation limit of 0.52 vol%. The low percolation limit, which indicated that graphene was well dispersed in the epoxy matrix, is an advantage of the 2D morphology. The EMI SE of the polymer/graphene composites (PGCs) was evaluated in the X-band range, as shown in Figure 4.12b. A maximum EMI SE of ~21 dB was achieved at a filler content of only 15 wt%, which meets the commercial requirement of 20 dB. It is important to note that an EMI SE of over 20 dB was obtained for a material with a low graphene content and superior mechanical properties, which is a clear advantage over PG films. Furthermore, the EMI SE can be increased with the addition of more filler to meet specific industrial requirements.

Magnetic nanoparticles can further improve the EMI shielding properties by providing dielectric permeability, which is otherwise not achievable in pure conducting materials. Thus, a variety of magnetic nanoparticles, including iron oxide, ferrites, and sendust, have been used to enhance the EMI SE of graphene–polymer composites. In one such effort, Dalal et al. developed laminated graphene and $PbTiO_3$-reinforced poly(3,4-ethylenedioxythiophene) (PEDOT) nanocomposites [56]. The polymer nanocomposites exhibited a core–shell structure containing $PbTiO_3$ and graphene as filler materials. The composites were prepared in situ via a chemical oxidative method with doping by 4-dodecylbenzenesulfonic acid (DBSA), which also acted as a surfactant. Polymerization was initiated by the addition of ammonium peroxydisulfate (APS), and the polymer layers grew on PbG

($PbTiO_3$ and RGO), leading to composites with a core–shell structure incorporating RGO and $PbTiO_3$ (Figure 4.12c). This architecture greatly enhanced the dielectric parameters that promote microwave absorption by the material.

The shielding properties were investigated in the 12–18 GHz range as a function of $PbTiO_3$ content by preparing different PPbG composite samples with PEDOT/RGO/$PbTiO_3$ weight ratios of 1 : 0.5 : 0 (PG), 1 : 0.5 : 1 (PPbG1), 1 : 0.5 : 2 (PPbG2), 1 : 0.5 : 3 (PPbG3), and 1 : 0.5 : 4 (PPbG4). Figure 4.12d shows that the absorption contribution to shielding increased with the frequency, whereas the reflection contribution decreased as the frequency increased. Enhanced absorption could be realized owing to the core–shell structure and the presence of $PbTiO_3$ and RGO in the polymer, which enhanced the dielectric parameters through interfacial and dipole polarization. The maximum SE_A at 18 GHz for the PPbG4 sample was 46.1 dB, whereas those of the PPbG1, PPbG2, and PPbG3 samples were 28.1, 35.0, and 40.1 dB, respectively, at 18 GHz and a thickness of 2.5 mm. The shielding properties were obviously enhanced by the addition of $PbTiO_3$, which provided a high dielectric loss contribution and corresponding increases in SE_A and SE_R (Figure 4.12e). Furthermore, the residual groups and defects in RGO also acted as polarized centers, resulting in an increased dielectric loss contribution. The PEDOT layer over the nanoparticles also provided good impedance matching with the space impedance, thus allowing more EMWs to enter the core–shell-structured material. The core–shell structure accumulated the space charge at the interface between PEDOT and the $PbTiO_3$/RGO composites, thus enhancing EMW absorption.

Despite the considerable progress that has been made in the synthesis of graphene–polymer composites, high filler contents and large thicknesses have motivated researchers to apply novel techniques to improve the percolation threshold and subsequently increase the overall EMI SE. To achieve this target, the formation of segregated structures by arranging graphene materials on the shell of polymer beads or particles is considered a viable approach for producing polymer composites to achieve the percolation threshold and hence good electrical conductivity and EMI SE at very low filler contents [57]. In this context, graphene was used to construct segregated conductive networks in an ultra-high-molecular-weight polyethylene (UHMWPE) matrix to realize an electrical conductivity of 0.04 S m^{-1} at a small filler content of 0.6 vol% [58]. For segregated composites consisting of UHMWPE containing 0.660 vol% (or 1.50 wt%) thermally reduced GO, an EMI SE of 28.3–32.4 dB was achieved in the X-band range at a thickness of 2.5 mm [59].

In another outstanding work, Yan et al. developed RGO/polystyrene (PS) segregated composites via high-pressure solid-phase compression molding [60]. In this configuration, the PS beads were added to a GO dispersion, and the GO was reduced in situ via hydrazine treatment. The product was filtered and compressed at a high pressure of 350 MPa to construct segregated polymer composites, as shown in Figure 4.13a–d. The authors compared the electrical and mechanical properties with conventional-pressure-assisted samples prepared at a low pressure of 5 MPa. Figure 4.13e shows the typical stress–strain curves of the two types of samples. Clearly, the high-pressure-compressed samples exhibited better mechanical properties than the conventional-pressure-assisted samples.

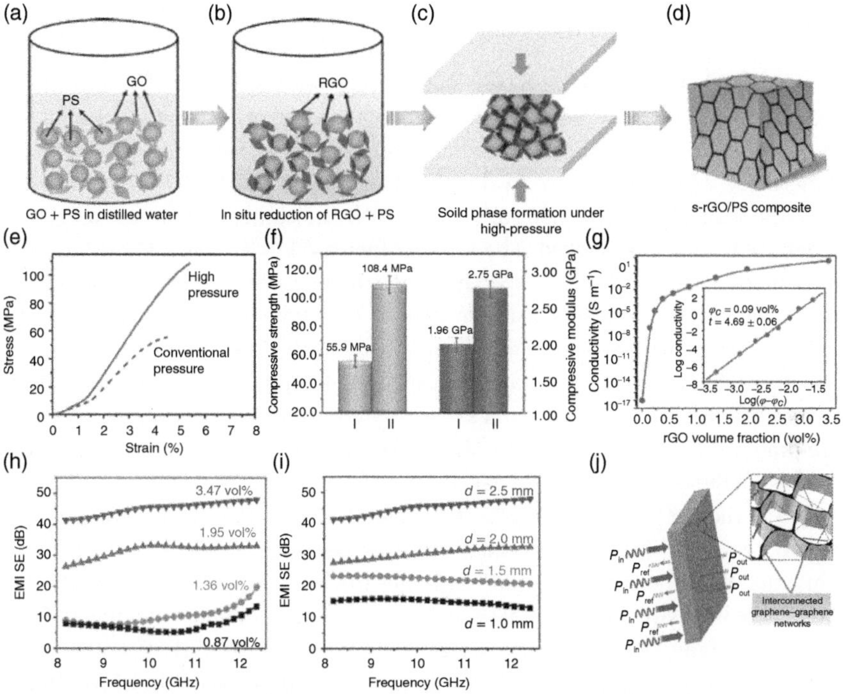

Figure 4.13 Schematic of the fabrication process of segregated RGO/PS (s-rGO/PS) composites: (a) GO and PS particles in distilled water, (b) in situ reduction of the GO/PS mixture in the presence of hydrazine hydrate, (c) RGO/PS powder after the removal of water, and (d) final s-rGO/PS composite. (e) Stress–strain curves of s-rGO/PS composites (1.95 vol% RGO) molded under high and conventional pressures. (f) Compressive strength and modulus of s-rGO/PS composites molded under conventional and high pressures (I and II, respectively). (g) Electrical conductivity of s-rGO/PS composites as a function of RGO loading. (h) EMI SE of the s-rGO/PS composites with various RGO loadings. (i) EMI SE of the 3.47 vol% s-rGO/PS composite at different thicknesses. (j) Schematic representation of microwave transfer across the s-rGO/PS composite. Source: Yan et al. [60]. Reproduced with permission of Wiley-VCH.

Figure 4.13f summarizes the compressive strength and modulus of both types of composites. The high-pressure-compressed composite exhibited an outstanding compressive strength (108 MPa) and modulus (2.75 GPa) as compared to the conventional-pressure-assisted sample (compressive strength of 55.9 MPa and modulus of 1.96 GPa). The enhanced mechanical properties in the case of high-pressure compressive molding were attributed to the interdiffusion of polymer chains across the boundaries between PS and the RGO flakes. Figure 4.13g shows the variation in electrical conductivity as a function of RGO content in the RGO/PS composite. The electrical conductivities of the composites were nearly 10 orders of magnitude higher than those of pristine PS and the RGO (0.14 vol%)/PS composite. The percolation threshold was 0.09 vol%, which is extremely low when compared to those of other nanomaterials such as graphite and carbon black [61, 62].

The extremely low percolation threshold of the RGO/PS composite was ascribed to the large aspect ratio of the RGO sheets and the dense conducting region formed around the PS spheres owing to the high compression pressure. The maximum conductivity of 43.5 S m^{-1} was found at an RGO loading of 3.47 vol%. The samples were subjected to EMI SE measurements, as shown in Figure 4.13h. The EMI SE increased with the increase in filler loading. The average EMI SE of the RGO composite with a loading of 1.36 vol% was 10.3 dB, indicating that more than 90% of incident radiation was blocked. The EMI SE increased to 31.5 and 45.1 dB at filler contents of 1.95 and 3.47 vol%, respectively, at a thickness of 2.5 mm. The increase in EMI SE was related to the increasing content of RGO, which improved the electrical conductivity and in turn the shielding properties. The thickness of the polymer composite is another crucial factor that dictates the shielding properties. To explore this, polymer composites of different thicknesses and a fixed filler content of 3.47 vol% were prepared (Figure 4.13i). The EMI SE was greatly improved by increasing the thickness of the composite. For instance, increasing the thickness from 1 to 2.5 mm resulted in the EMI SE increasing from 15.2 to 41.4 dB at 8.2 GHz and from 12.9 to 48 dB at 12.4 GHz. This enhancement occurred because more RGO was available to interact with the incoming EMWs in the thicker materials. Importantly, the SE depended heavily on the PS particle size. At a fixed amount of graphene filler (3.47 vol%), a reduction in PS particle size led to a decrease in EMI SE, whereas the EM SE increased when the PS particle size increased from 600 nm to 96 μm. In the small PS particles, the surface area of insulating PS increased, whereas the graphene sheets interacted with each spontaneously, giving rise to a thin anisotropic film without 3D electrical conductive channels. In the large PS particles, the graphene sheets could surround the PS particles to create a 3D percolative structure, thus providing larger conductive channels for EMW mitigation. The segregated polymer composites showed an absorption-dominant shielding mechanism, as the internal facets provided multiple reflections that promoted EMW absorption in the material. The shielding mechanism is illustrated in Figure 4.13j. The incident EMWs entering the segregated structure were attenuated by reflection, scattering, and absorption. A portion of the incident waves was reflected owing to the conductive nature of the material, whereas the remaining EMWs underwent multiple reflections inside the material owing to the multiple facets of PS.

In another report, Zhang et al. developed a functional PMMA/graphene nanocomposite microcellular foam by blending PMMA and graphene sheets, followed by a batch foaming process with subcritical CO_2 gas [63]. The bulk composite samples were saturated with CO_2 under experimental conditions (0–25 °C, 3.5–5.0 MPa, and 24 hours) in a pressure vessel. After complete saturation, the pressure was rapidly released and the samples were immersed in preheated hot water for a specific time to produce samples with various densities and graphene contents. Figure 4.14a shows the stress–strain curves of neat PMMA and the PMMA/graphene foam (GF) composites. The addition of graphene sheets made the composites brittle, as evidenced by the reduced ductility and tensile toughness; however, the ductility of the PMMA/graphene composite foam samples (curves 5 and 6) significantly increased after the foaming process. For example, the PMMA foam with 0.5 wt% graphene sheets exhibited a fracture strain of 24%, whereas that of its bulk counterpart (PMMA/graphene) was 5%. Figure 4.14b shows the variation in electrical conductivity as a function of graphene loading. The foam composites exhibited better electrical conductivity with the addition of graphene, and a good conductivity value of 3.11 S m^{-1} was achieved for the foam composite with a graphene content of only 1.8 vol%. Figure 4.14c shows the EMI SE in the X-band range for different samples. The highest EMI SE of 19 dB was obtained for the foam composite with 1.8 vol% graphene at a thickness of 2.4 mm, which is superior to the shielding performance of many commonly used bulk polymer composites. Figure 4.14d shows the contributions from SE_A and SE_R. Almost all the incoming EMWs were absorbed in the microcellular foam composite but the contribution from reflection was negligible. In addition to the reflection and absorption contributions, multiple internal reflections also played a significant role in enhancing the overall EMI SE. The microcellular structure of the PMMA foam composites provided a large cell surface and hence a larger surface area for interactions between the EMWs and the graphene sheets on the periphery of the cellular structure. The incident EMWs entering the PMMA/GF were reflected and scattered many times inside the material, resulting in energy loss.

Zheng and coworkers also developed a low-density and compressible PGC foam for EMI SE [64]. A polyurethane/graphene (PUG) foam was fabricated by a simple dip-drying strategy, as shown in Figure 4.14e–g. The commercial polyurethane (PU) sponge was dip-coated in the as-prepared GO solution at a concentration of 3 mg ml^{-1} by repetitive squeezing and drying in an air-circulating oven at 90 °C. The graphene-coated PU was subjected to a hydrothermal reaction with hydrazine monohydrate to reduce GO. The loading of graphene on PU was determined by weighing the PU foams before and after the coating process. Figure 4.14h shows the EMI SE of PUG-10 (10 wt% graphene content) in the X-band range. Upon the application of a compressive strain from ~0% to 25%, 50%, and 75%, the EMI SE of PUG-10 with a thickness of ~6 cm was reduced from 39.4 to 34.9, 26.2, and 23.4 dB, respectively. The decrease in EMI SE was mainly attributed to a reduced contribution from SE_A because less of the foam structure was available for EMW interaction, whereas there was negligible variation in the SE_R contribution. Although the electrical conductivity of PUG-10 slightly increased as a result of compressive strain, the surface area for EMW interaction was significantly reduced, which minimized the role of multiple internal reflections and resulted in a lower SE. Because the surface did not change, the impedance matching remained the same, and therefore, the reflection contribution was not significantly affected.

Wu et al. utilized a different approach for creating foam structures by depositing graphene on Ni foam, followed by etching of Ni and drop-coating of PEDOT:PSS (poly(3,4-ethylenedioxythiophene) polystyrene sulfonate). The Ni foam was initially heated in a CVD system at 1000 °C, followed by the introduction of CH_4 under argon [65]. Freestanding GFs were obtained by etching the Ni metal. Then, the GFs were noncovalently functionalized with DBSA to improve the wettability and enhance the adhesion between GF and PEDOT:PSS. The functionalized GFs were

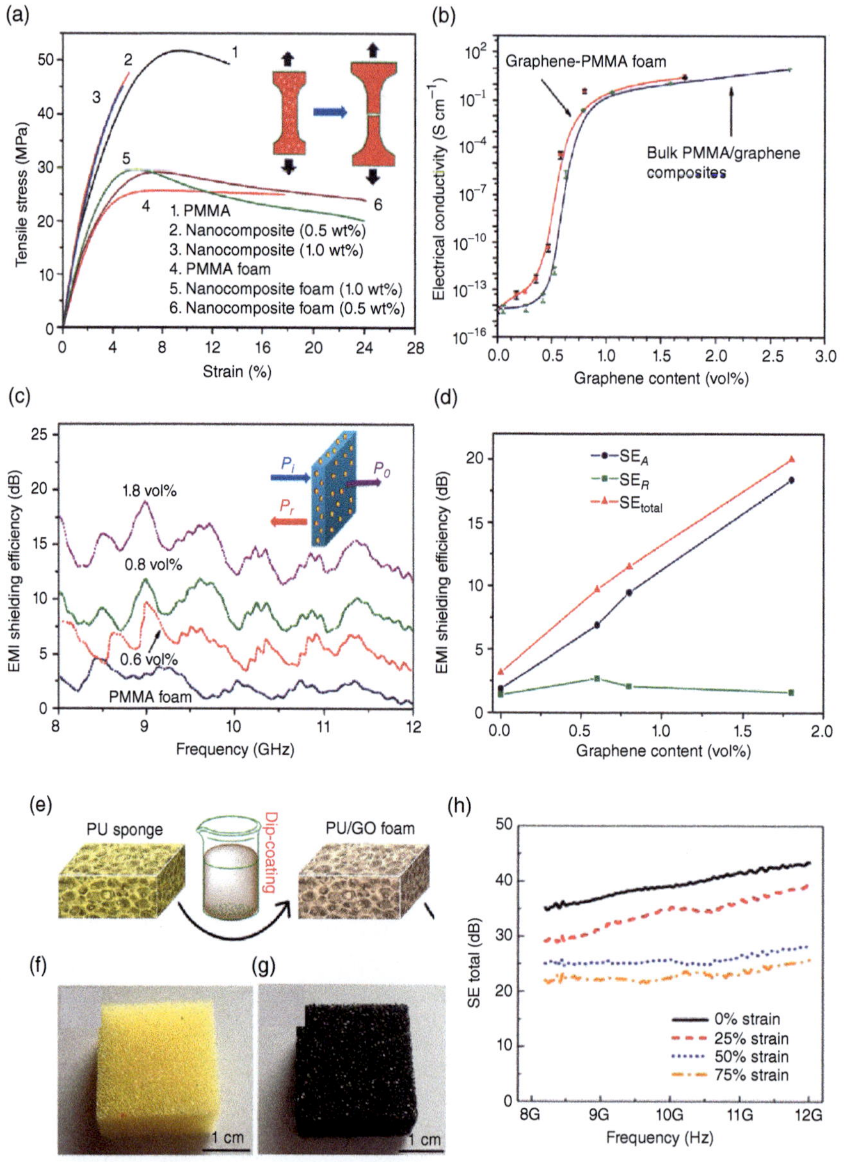

Figure 4.14 (a) Stress–strain curves of PMMA and its graphene nanocomposites before and after foaming, (b) electrical conductivity vs. graphene content for bulk PMMA/graphene nanocomposites and microcellular foams, (c) EMI SE of PMMA/graphene nanocomposite microcellular foams with different graphene sheet contents, (d) comparison of SE_T, microwave absorption (SE_A), and microwave reflection (SE_R) at 9 GHz. Source: Zhang et al. [63]. Reproduced with permission of the American Chemical Society. (e–g) Fabrication process of PUG foams, including dip-coating of GO sheets onto a PU framework, and subsequent reduction to yield a black color. (h) SE_T of the PUG-10 foam at a thickness of ~6 cm under different compressive strains. Source: Shen et al. [64]. Reproduced with permission of the American Chemical Society.

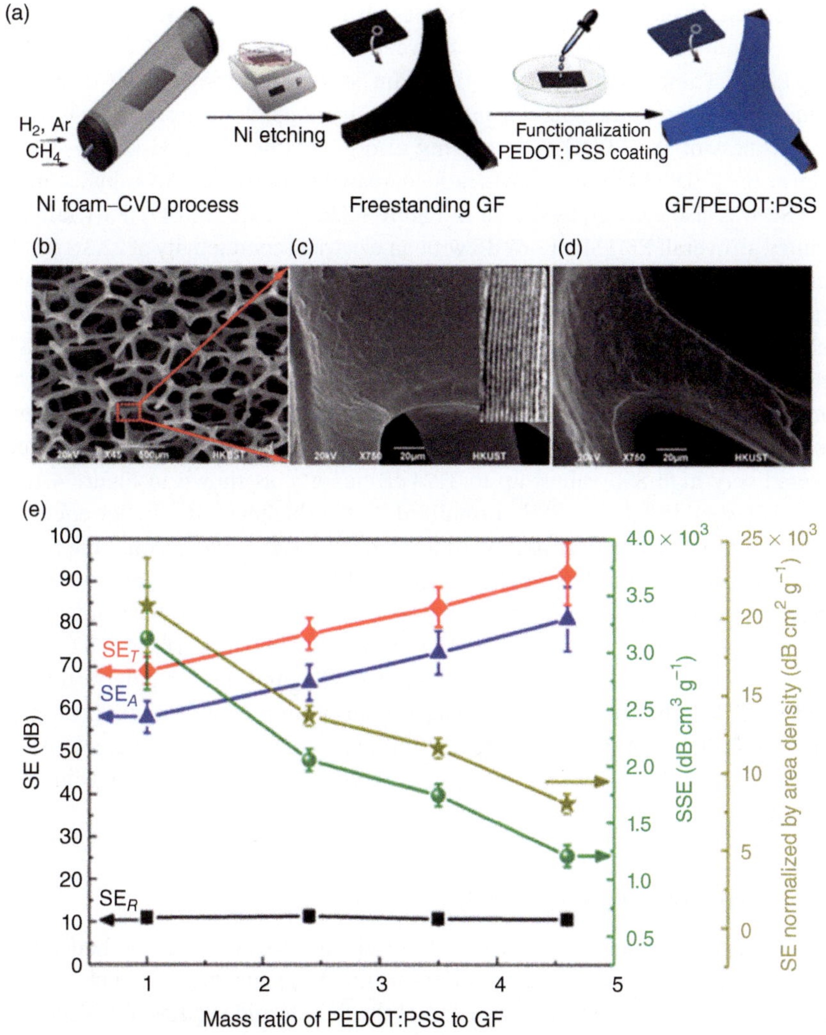

Figure 4.15 (a) Schematic for the preparation of GF/PEDOT:PSS composites. SEM and TEM images of GFs (b, c) before and (d) after PEDOT:PSS coating. (e) EMI SE, SSE, and SSE/*t* as a function of the PEDOT:PSS to GF mass ratio. Source: Wu et al. [65]. Reproduced with permission of the American Chemical Society.

drop-coated with a mixture of PEDOT:PSS and DMSO (dimethyl sulfoxide) to obtain different mass ratios of the polymeric material (Figure 4.15a). Figure 4.15b shows the morphology of the freestanding GFs, which consisted of a cellular porous structure with wrinkles on the surface arising from the grain boundaries of the underlying Ni foam template. Figure 4.15c shows that the prepared GF was composed of 12 layers of graphene. Figure 4.15d shows the uniform coating of PEDOT:PSS on the GF. The electrical conductivity of the GF/PEDOT:PSS foam gradually increased with an increase in the PEDOT:PSS mass content, reaching a maximum value of 43.2 $S\,cm^{-1}$ at a PEDOT:PSS to GF mass ratio of ~7. The improvement in electrical conductivity was ascribed to the connection of large openings, both internal and external, between the skeleton of the GF foams and PEDOT:PSS. These connections in the foam structure improved the electrical conductivity owing to the superior intrinsic electrical conductivity of PEDOT:PSS. The thickness of the PEDOT:PSS coating also gradually increased from ~0.8 to 3.9 μm as thc PEDOT:PSS to GF mass ratio increased from 0.7 to 7. An enhancement of the SE was also evident, as shown in Figure 4.15e. The GF/PEDOT:PS-1 sample exhibited an overall EMI SE of ~69 dB with an electrical conductivity of 22.3 S cm^{-1}, whereas GF/PEDOT:PS-4.6 exhibited an EMI SE of 91.9 dB with an electrical conductivity of 35.2 S cm^{-1}. These results revealed that absorption was the dominant shielding mechanism. As foam structures are generally lightweight, the EMI SSE of these structures, obtained by normalizing SE_T by the composite density, was typically superior to that of the bulk counterparts. The GF/PEDOT:PS composites possessed very low apparent densities ranging from 0.0221 to 0.0762 g cm^{-3}, which delivered very high SSE values up to 3124 dB cm^3 g^{-1}, as shown in Figure 4.15e. SSE/t, corresponding to the SSE normalized by the thickness, is another effective way to describe the shielding ability. At a constant thickness of 1.5 mm, the SSE/t for the composites ranged from 8040 to 20 800 dB cm^2 g^{-1}, which are among the highest values of all materials produced to date. The shielding mechanisms in these composite samples were ascribed to the conducting network, which caused interfacial polarization between GF and PEDOT:PSS, and dipolar polarization in the PEDOT:PSS chains, which improved the absorption contribution. Thus, the capacitor-like behavior induced by polarization and the resistor-like characteristics originating from the conductive composite network were mainly responsible for shielding.

4.2.4 Heteroatom-Doped Graphene

Conventionally, graphene is synthesized via chemical exfoliation methods. The harsh chemical reaction generates defects in the graphene sheets, which impair the electrical properties. To heal a portion of the defects and improve the electrical properties, graphene can be doped with n-type dopants that fill the vacant sites and provide extra electrons for electronic conduction. Doping with heteroatoms such as sulfur, nitrogen, boron, fluorine, and phosphorus has been successfully demonstrated for various applications. Among these heteroatom dopants, sulfur and nitrogen have been shown to provide n-type doping, which not only increases the electrical conductivity but also enhances the EMI SE. Shahzad et al. reported the first-ever sulfur-doped graphene for EMI shielding applications. The authors varied the doping reaction temperature to control the level of doping as well as the degree of GO reduction [66]. Sulfur-doped reduced graphene oxide (SrGO) was synthesized via thermal-driven high-temperature doping reactions at 250, 650, or 1000 °C under H_2S gas flow for 10 or 30 minutes. The samples were maintained at the same temperature for another 30–50 minutes for annealing. The different temperatures and times were used to control the dopant content in graphene. To investigate the EMI shielding properties, the S-doped and undoped samples

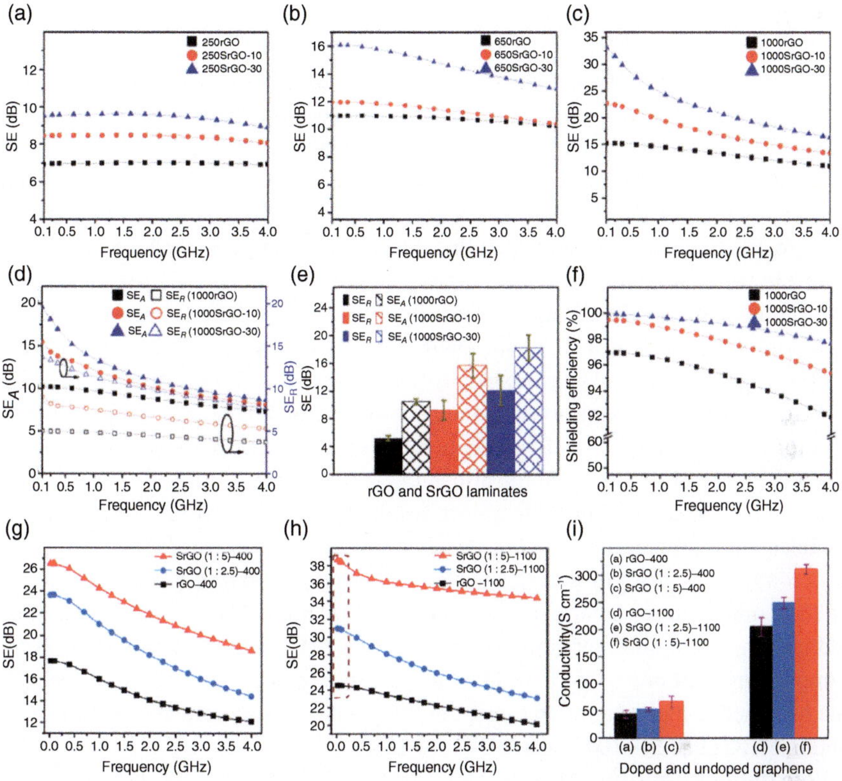

Figure 4.16 EMI SE of (a) 250rGO, 250SrGO-10, and 250SrGO-30; (b) 650rGO, 650SrGO-10, and 650SrGO-30; and (c) 1000rGO, 1000SrGO-10, and 1000SrGO-30. (d) SE_A and SE_R of 1000rGO, 1000SrGO-10, and 1000SrGO-30. (e) Comparison of SE_A and SE_R for 1000rGO, 1000SrGO-10, and 1000SrGO-30 at 100 MHz. (f) Shielding efficiency (%) for 1000rGO, 1000SrGO-10, and 1000SrGO-30 laminates as a function of frequency. Source: Shahzad et al. [66]. Reproduced with permission of the Royal Society of Chemistry. EMI SE of doped and undoped graphene samples prepared via low- and high-temperature lenthionine reactions: (g) rGO-400, SrGO (1 : 2.5)-400, and SrGO (1 : 5)-400 and (h) rGO-1100, SrGO (1 : 2.5)-1100, and SrGO (1 : 5)-1100. (i) Electrical conductivity of doped and undoped graphene laminates. Source: Shahzad et al. [67]. Reproduced with permission of the American Chemical Society.

were compressed to form laminates. Figure 4.16a–c illustrates the EMI SE of the as-synthesized SrGO and RGO laminates at different temperatures and doping times. The sample designations indicate the preparation conditions, for example, 1000SrGO-30 represents the sample doped at 1000 °C for 30 minutes and 1000rGO represents undoped RGO annealed for one hour. The shielding results clearly showed that doping played a significant role in enhancing the EMI SE of the RGO laminates. Increasing the doping content, as indicated by the reaction time, also enhanced the EMI SE.

The largest EMI SE value of 33.2 dB at 100 MHz was obtained for 1000SrGO-30, which is 119% larger than that of the undoped sample (15.5 dB) with the same thickness of 0.14 mm. Figure 4.16d shows the contributions of reflection and absorption to the EMI SE. For all the samples, absorption was the dominant shielding mechanism. The difference between the absorption and reflection contributions is shown in Figure 4.16e for the 1000 °C samples at a frequency of 100 MHz. The doped samples treated at higher temperatures and larger doping times displayed better shielding results. In all cases, the SE_A was higher than the SE_R. Furthermore, the shielding efficiency (%) was also determined for the 1000 °C samples, where 1000SrGO-30 provided >99.9% blockage of the incident radiation at 100 MHz (Figure 4.16f), whereas poorer shielding performance was observed for the sample with a lower S-content.

In another report, Shahzad et al. developed n-type, S-doped graphene using a lenthionine-based precursor as an environment-friendly sulfur source [67]. The synthesis process involved a two-step procedure, in which GO was first pre-reduced to RGO at a lower temperature to avoid material loss during the exothermic reaction and for safety reasons. The doping content was controlled by the sulfur source in the precursor material as well the annealing temperature. A higher amount of the sulfur source and a higher temperature resulted in better EMI SE properties. The samples were prepared by mixing the pre-reduced RGO powder and lenthionine (1 : 2.5 and 1 : 5, w/w) in a vortex shaker, followed by heating to 400 °C at a rate of 10 °C min^{-1} under an argon atmosphere for one hour. The sample designations indicate the preparation conditions. For example, SrGO (1 : 2.5)-400 represents the sample treated at 400 °C with an RGO to lenthionine ratio of 1 : 2.5. To investigate the behavior at higher temperatures, the samples were also annealed at 1100 °C to obtain SrGO (1 : 2.5)-1100 and SrGO (1 : 5)-1100.

Both the doped and undoped samples were compressed in custom-designed molds to fabricate thin laminates for determining the EMI SE properties. Figure 4.16g,h shows the EMI SE of rGO-400, SrGO (1 : 2.5)-400, and SrGO (1 : 5)-400 with values of 17.6, 23.6, and 26.5 dB, respectively, at 25 MHz and values of 12.0, 14.3, and 18.6 dB, respectively, at 4 GHz at a thickness of 0.15 mm. A significant increase in the EMI SE was observed after the high-temperature annealing process, with EMI SE values of 24.4, 31.0, and 38.6 dB observed at 25 MHz for rGO-1100, SrGO (1 : 2.5)-1100, and SrGO (1 : 5)-1100, respectively, at a thickness of 0.15 mm (Figure 4.16h). The heavily doped graphene laminate SrGO (1 : 5)-1100 exhibited a 58% increase in the EMI SE value compared with that of the rGO-1100 laminate. In addition, SrGO (1 : 5)-1100 exhibited an almost 50% increase in electrical conductivity (311 S cm^{-1}) as compared to undoped rGO-1100 (205 S cm^{-1}), as shown in Figure 4.16i. The SE_A and SE_R or SrGO (1 : 5)-1100 were superior to those of the other samples, and the absorption contribution was found to be greater than the reflection contribution in all the samples. Other factors such as the laminate form also played an important role in the EMI SE, as shielding by absorption was further promoted by multiple internal reflections. Furthermore, the surface functional groups acted as polarization centers, where EMWs can be blocked because of energy loss through interactions with the surface groups.

In another work, Umrao et al. utilized a microwave-assisted approach to develop reduced graphene oxide (MRG), boron-doped reduced graphene oxide (B-MRG), nitrogen-doped reduced graphene oxide (N-MRG), and B,N-codoped reduced

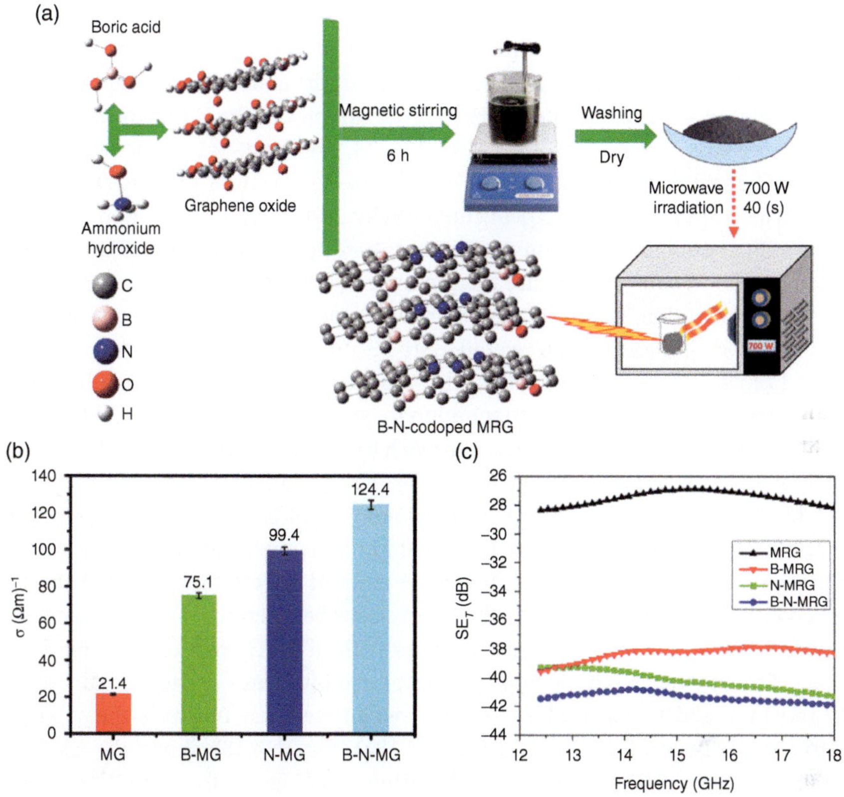

Figure 4.17 (a) Microwave-assisted method for the preparation of B,N-codoped MRG (reduced graphene oxide). (b) Electrical conductivities of MRG, B-MRG, N-MRG, and B-N-MRG at room temperature. (c) EMI SE of MRG, B-MRG, N-MRG, and B-N-MRG in the Ku-band range. Source: Umrao et al. [68]. Reproduced with permission of the American Chemical Society.

graphene oxide (B-N-MRG) [68]. The doped samples were prepared by the simple chemical mixing of precursors, such as boric acid (for boron doping) and ammonia solution (for nitrogen doping), with GO at 60 °C for eight hours. After drying the mixture at 80 °C, the obtained solid product was exfoliated using a microwave oven at 700 W for 40 seconds, as shown in Figure 4.17a. The microwave-assisted approach is an attractive synthesis process because it offers ease of processability, short reaction times, and high yields; moreover, it does not require high temperatures of hazardous reductants. It is well known that N doping results in an n-type doping effect by providing more electrons to the graphene lattice, whereas B doping creates holes in the graphene sheet [69, 70]. Both these methods promote electron transfer between the valence and conduction bands and hence affect the electrical conductivity.

Figure 4.17b shows the electrical conductivity of the samples, as measured using the four-probe conductivity method. The room-temperature conductivity of the samples increased from 21.4 S m^{-1} for MRG to 124.4 S m^{-1} for B-N-MRG. N and B codoping enhanced the conductivity by introducing free electrons and holes into the graphene π-conjugation system, which may create a junction within the graphene layer. The samples exhibited nonlinear variations in resistance with temperature, which was explained by thermally activated transport and variable-range hopping models. Figure 4.17c shows the EMI SE of the samples in the Ku-band range. The EMI SE increased from −28 dB for MRG to −38, −40, and − 42 dB for B-, N-, and B-N-MRG at a thickness of 1.2 mm. The increased EMI SE of the doped samples was ascribed to their high electrical conductivity. A detailed investigation revealed that the absorption contribution was greater than the reflection contribution. SE_R remained almost constant because the presence of both B and N in MRG enhanced the spin polarization, space charge polarization, and natural resonances, leading to high absorption and a high overall EMI SE. Furthermore, the doped samples showed much lower skin depths than the undoped samples, and this shallow skin depth allowed the same level of attenuation to be obtained with a thinner shield.

Chhetri et al. utilized a synergistic approach to increase the EMI SE by anchoring Fe_3O_4 nanoparticles on N-doped RGO [71]. Fe_3O_4 nanoparticles were prepared by a hydrothermal route, whereas GO reduction was performed with hydrazine monohydrate. Fe_3O_4@N-doped-rGO hybrid epoxy composites were prepared by solution processing, which anchored a large amount of Fe_3O_4 on the RGO sheets. The coordination bonds formed by the interactions between the lone-pair electrons of N and the d-orbitals of Fe assisted the growth of Fe_3O_4 particles on the RGO sheets. The Fe_3O_4@N-rGO hybrid composite exhibited an electrical conductivity of 1.08 S m^{-1}, which increased to 1.22 S m^{-1} when the filler content in the polymer matrix was between 3 and 5 wt% (Figure 4.18a). Figure 4.18b shows the EMI SE of the polymer hybrid composites with various loadings in the 2–18 GHz range. The EMI SE increased with the increase in filler content, and the maximum EMI SE of ~26 dB was observed for 15 wt% filler at a thickness of 1 mm. The sample exhibited frequency dependence with a greater absorption contribution resulting from the presence of magnetic nanoparticles. The reflection contribution remained nearly constant but decreased slightly with the increase in frequency. The increased absorption contribution was shown to originate from the increase in interfacial polarization resulting from the larger electric and magnetic dipoles caused by the presence of magnetic Fe_3O_4. The anchoring of Fe_3O_4 on RGO promoted interfacial polarization and magnetic losses, which provided the necessary impedance-matching condition for EMW absorption [73].

Wan et al. used large graphene sheets and a doping strategy to develop high-performance EMI shielding materials [72]. Three different types of graphene sheets were prepared, namely, SGO , MGO (medium graphene oxide), and LGO. GO papers were prepared from SGO, MGO, and LGO (denoted SG, MG, and LG, respectively) by vacuum filtration assembly, followed by high-temperature graphitization treatment between graphite plates at 1600 °C. The LG paper was doped with iodine by exposure to iodine vapor at 200 °C for 12 hours to fabricate iodine-doped LG (I-LG). Excess iodine was removed by placing the I-LG paper in a vacuum oven at 60 °C for two hours.

The EMI SE performance of the samples was evaluated in the X-band range, as shown in Figure 4.18c,d. The EMI SE of MG with a thickness of 8.4 μm at

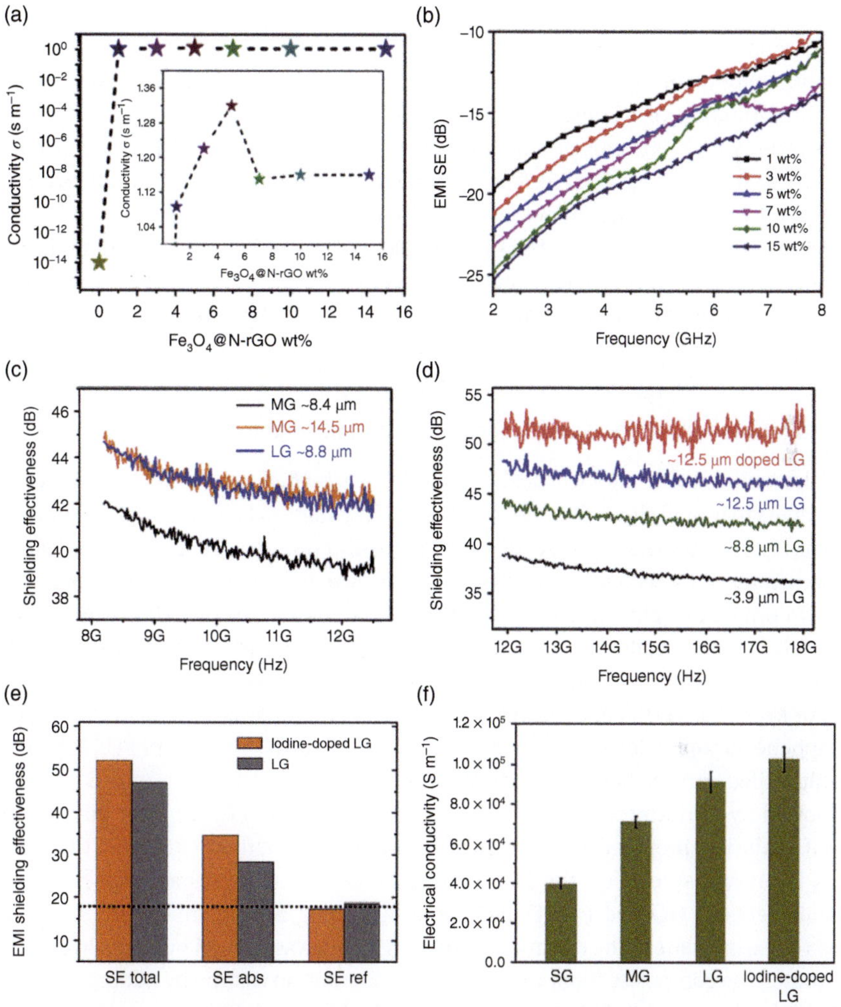

Figure 4.18 (a) Electrical conductivity and (b) EMI SE of Fe_3O_4@N-rGO/epoxy composites. Source: Chhetri et al. [71]. Reproduced with permission of Elsevier. (c) EMI SE of graphene paper samples with different sheet sizes. (d) EMI SE of LG with different thicknesses and the improvement after iodine doping. (e) EMI SE, SE_A, and SE_R and (f) electrical conductivity of the graphene papers. Source: Wan et al. [72]. Reproduced with permission of Elsevier.

8.2 GHz was 42 dB, whereas that of LG with a similar thickness increased to 44.7 dB. Interestingly, the EMI SE of LG was equivalent to that of an MG paper with almost twice the thickness (14.5 μm). The EMI SE of graphene papers increased with an increase in thickness. The EMI SE of LG increased to 65.3 dB as the thickness increased to 21.3 μm, which was much higher than the EMI SE of SG at a comparable thickness (55.2 dB at 21.2 μm). The enhanced EMI SE was ascribed to LG sheets having fewer defects and more conjugated carbon domains than SG sheets, which resulted in higher mobility. The I-LG paper also displayed an outstanding EMI SE of 52.2 dB at a thickness of 12.5 μm (Figure 4.18d). A detailed investigation revealed an absorption-dominant EMI shielding phenomenon (Figure 4.18e). The good EMI SE performance mainly originated from the better electrical conductivity of the graphene paper as well as the doping. Iodine doping results in the formation of triiodide (I_3^-) and pentaiodide (I_5^-) through charge transfer processes, which can substantially improve the carrier density of I-LG, thus leading to higher conductivity [74, 75]. The SG paper displayed an electrical conductivity of ~3.94×10^4 S m^{-1}, which increased to 9.12×10^4 S m^{-1} for LG and 1.05×10^5 S m^{-1} for I-LG (Figure 4.18f). The I-LG paper also had a lower skin depth than all the other samples, which means that the exponential decrease in electromagnetic energy occurred at a much smaller thickness in I-LG than in the other samples, thus improving the overall EMI SE.

4.2.5 Graphene Hybrids with Other Carbon Materials

PG has an outstanding electrical conductivity, which makes it a front runner in the quest to find high-performance EMI shielding materials. However, conductivity alone is not responsible for the high EMI SE of different materials. The polarization losses induced by secondary fillers or the presence of other fillers, magnetic losses, and internal multiple reflections also play crucial roles in the overall EMI SE. Thus, hybrid composite materials based on graphene and other carbon materials have been extensively researched to enhance the EMI shielding properties.

For example, Verma et al. developed a ternary hybrid nanocomposite based on a graphene nanoplatelet–carbon nanotube (GCNT) hybrid as a filler in PU [76]. First, multiwalled carbon nanotubes (MWCNTs) were synthesized via a CVD process, followed by acid functionalization to obtain FCNTs. The FCNTs were mixed with a GO solution prepared by the common Hummer's method (Figure 4.19a). The π–π interactions between the FCNTs and the GO basal planes and the hydrogen bonds between GO and FCNTs facilitated the strong absorption of the FCNTs on the GO basal planes. The chemical reduction of GO was carried out with hydrazine hydrate. The composite films were prepared via solution casting by adding different GCNT loadings (0.5, 1, 2, 3, 5, and 10 wt%) in the PU matrix. The composite samples exhibited good electrical conductivities, as shown in Figure 4.19b. The conductivity of neat PU was 3.9×10^{-11} S m^{-1}, which increased to 9.5 S m^{-1} for the PUGCNT 10 nanocomposites. The FCNTs acted as a bridge between the graphene sheets, which increased the electrical conductivity of the hybrid. All the samples were subjected to EMI SE characterization (Figure 4.19c). Pure PU showed a negligible EMI SE owing to its insulating nature, whereas the EMI SE of the composites improved with increasing hybrid filler content, reaching −47 dB for the PUGCNT 10 nanocomposite. (It should be noted that the negative SE values reported in this work are equivalent to positive SE values. The negative sign originates from reversing the logarithm of the intensities of incident and transmitted EMWs.) In contrast, a physically mixed hybrid sample containing a 1 : 1 ratio of RGO and CNT with a loading of 10 wt% in PU (PURGOCNT) had an EMI SE of −32 dB, which is significantly lower than that of the PUGCNT 10 sample at a thickness of 3 mm. This marked difference was attributed to the synergistic effect originating from the

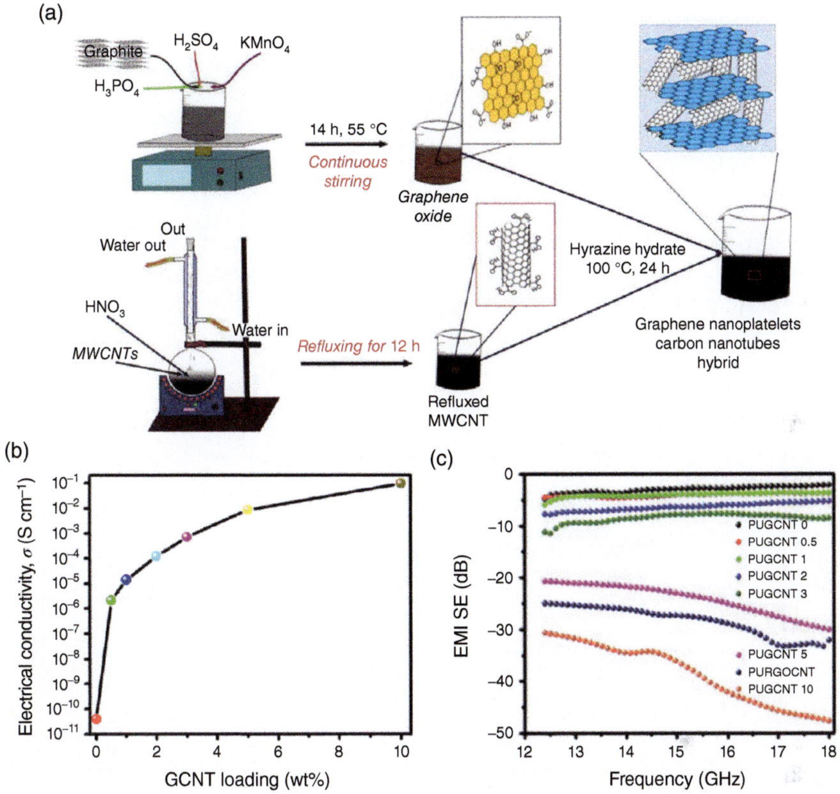

Figure 4.19 (a) Schematic representation of the synthesis of GCNT hybrids, (b) effect of GCNT content on the electrical conductivity of PU composites, and (c) variation in EMI SE with frequency for PUGCNT nanocomposites. Source: Verma et al. [76]. Reproduced with permission of Elsevier.

hybridization of the graphene sheets and FCNTs, with the FCNTs playing a crucial role as bridges to provide better channels for EMW attenuation. In a heterogeneous system, the accumulation of virtual charges at the interface between two media with different dielectric constants and conductivities leads to interfacial polarization, which greatly enhances the absorption of EMWs [77].

In another approach, Kong et al. used a catalytic growth procedure, in which CNTs were formed in situ on RGO sheets at 600 °C and filled into the polydimethylsiloxane (PDMS) to construct high-performance EMI shielding polymer composites [78]. First, GO and cobalt acetate solutions were mixed, sonicated, and freeze-dried to form an aerogel, which was subsequently heated in a horizontal tube furnace to reduce GO and the cobalt catalyst under a H_2/Ar atmosphere. The catalytic growth of CNTs was performed by flowing a H_2/Ar gas mixture through the liquid precursor (acetone) in the furnace at ambient pressure (Figure 4.20a). The purified CNT/G hybrid was mixed with a PDMS solution and cured at 130 °C to obtain flexible polymer composites.

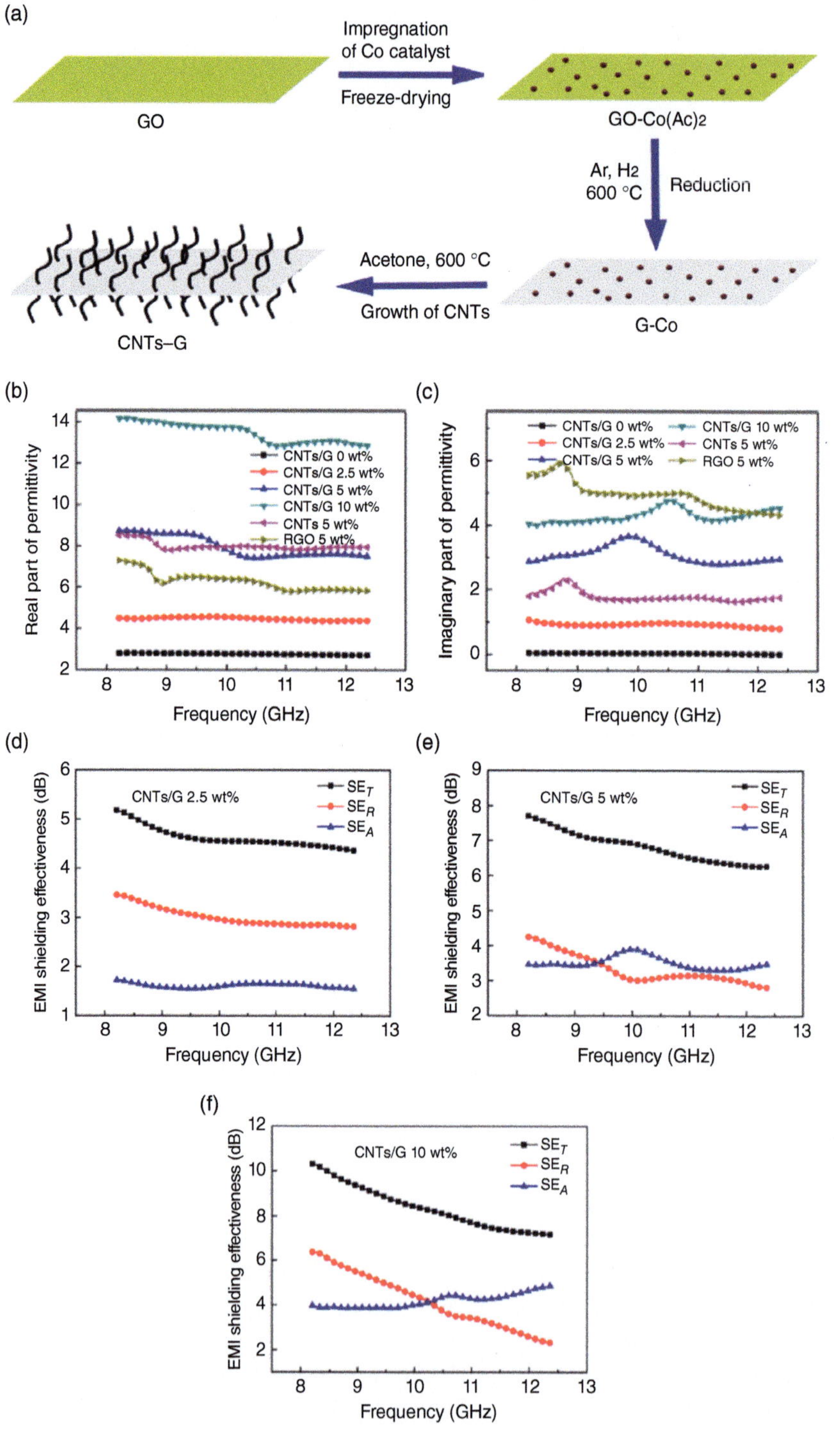

Figure 4.20 (a) Schematic of the synthetic process of CNT/G hybrids. (b) Real and (c) imaginary parts of the relative complex permittivity of CNT/G hybrids, RGO, and CNTs dispersed in PDMS. EMI SE of CNT/G–PDMS composites with loadings of (d) 2.5, (e) 5, and (f) 10 wt%. Source: Kong et al. [78]. Reproduced with permission of Elsevier.

Figure 4.20b,c shows the real and imaginary parts (ε', ε'') of the permittivity for polymer composites with different filler loadings. The ε' value increased with the increase in filler loading, with the 10 wt% CNT/G hybrid composite showing the maximum dielectric constant. This was because increasing the amount of CNT/G sheets improved the dipolar polarization and the electrical conductivity, so that more energy could be stored and in turn lost inside the material. Furthermore, the imaginary part (ε'') exhibited nonlinear behavior, with the highest value observed for 5 wt% RGO. The plot of ε' vs. ε'' (Cole–Cole plot) showed two semicircles for the 10 wt% CNT/G hybrid composite, whereas only one semicircle was observed for the 2.5 wt% CNT/G hybrid composite. Thus, the hybrid with the higher filler content had more Debye relaxation phenomena, which were responsible for the enhanced dielectric properties of the polymer composites. Figure 4.20d–f shows SE_T, SE_A, and SE_R for the CNT/G hybrid composites. The shielding properties were enhanced with the increase in filler content. The 10 wt% CNT/G hybrid composite exhibited the maximum EMI SE_T of 10.4 dB, which was two times higher than that of the 2.5 wt% CNT/G hybrid composite. The SE_A of the composites also increased with the increase in hybrid filler content. The enhanced shielding properties were mainly ascribed to the good electrical conductivity and the dipolar polarization induced by the addition of the CNT/G hybrid filler.

Huangfu et al. fabricated polymer composites based on a MWCNT/thermally annealed graphene aerogel (TAGA) with a polyaniline (PANI) matrix [79]. GO and the functionalized MWCNTs were dispersed with the addition of L-ascorbic acid to form a hydrogel, which was then frozen in liquid nitrogen and thermally annealed at 800 °C for 30 minutes to obtain MWCNT/TAGA. Subsequently, MWCNT/TAGA was placed in a solution containing a certain amount of aniline, and APS was slowly added in an ice bath for a duration of over one hour to obtain the PANI/MWCNT/TAGA composite (Figure 4.21a). The chemical for epoxy curing was added to the MWCNT/TAGA and PANI/MWCNT/TAGA mixtures to fabricate MWCNT/TAGA/epoxy and PANI/MWCNT/TAGA/epoxy nanocomposites. The samples with 0, 0.41, 0.83, and 1.24 wt% MWCNT filler, no PANI, and 1.20 wt% of TAGA were designated as CGE1, CGE2, CGE3, and CGE4, respectively, whereas those with fixed TAGA and MWCNT filler contents of 1.20 and 0.83 wt%, respectively, and 0, 1.34, 1.96, and 2.58 wt% PANI were designated as PCGE1, PCGE2, PCGE3, and PCGE4, respectively. Figure 4.21b,c shows the electrical conductivity of all the samples. The electrical conductivity increased with increasing MWCNT content. However, the sample with the highest MWCNT content (CGE4) exhibited a lower electrical conductivity owing to the agglomeration of MWCNTs. The electrical conductivities of the composites increased with the addition of the PANI matrix. When the PANI loading was 2.58 wt%, the electrical conductivity was improved from 42 to 52.1 $S\,m^{-1}$, mainly because of the conductive network of the PCGE composites. Figure 4.21d,e illustrates the EMI SE of all the samples including pure epoxy. The EMI SE of the CGE nanocomposite increased from 28 to 36 dB upon the addition of more MWCNTs at a thickness of 3 mm. However, similar to electrical conductivity, the EMI SE of CGE4 decreased slightly. The composites with PANI showed higher EMI SE values owing to better electrical conductivity and additional

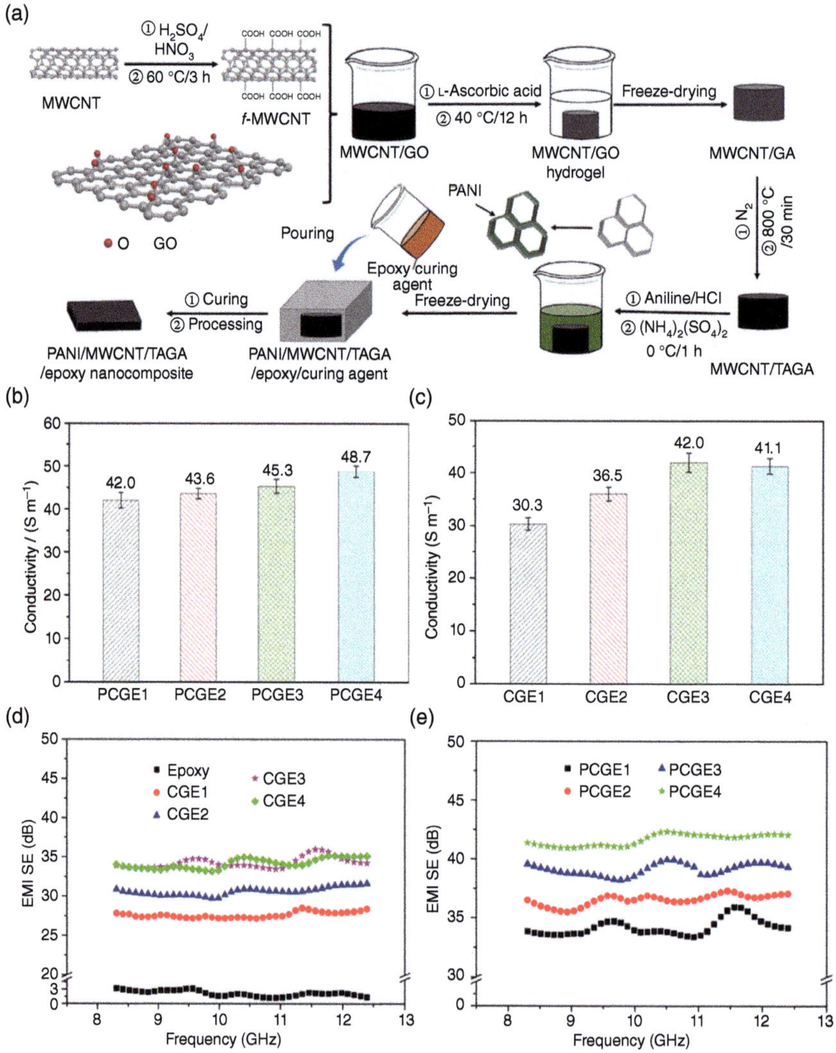

Figure 4.21 (a) Schematic of the fabrication of PANI/MWCNT/TAGA/epoxy nanocomposites. Electrical conductivities of (b) CGE and (c) PCGE nanocomposites. EMI SE of (d) CGE and (e) PCGE nanocomposites. Source: Huangfu et al. [79]. Reproduced with permission of Elsevier.

polarization effects. The SE_R of the PCGE nanocomposites increased from 5 to 7 dB, whereas the SE_A increased from 31 to 35 dB. The impedance mismatch between the PCGE nanocomposites and air gave rise to enhanced reflection. Furthermore, the improved electrical conductivity promoted eddy current and ohmic losses, which increased the SE_A contribution.

Song et al. developed an interesting approach for interfacial engineering by constructing a carbon nanofiber–graphene–carbon nanofiber (CNF–GN–CNF) heterojunction for flexible and lightweight EMI shielding materials [80]. A GN/CNF composite network was fabricated using an electrospinning technique. In this process, polyacrylonitrile (PAN) was added to a GN and DMF (dimethylformamide) suspension, and the mixture was loaded into a syringe with a stainless steel nozzle. A high voltage of ~15 kV was applied between the nozzle tip and the collector, and electrospinning was carried out at ambient temperature. The as-spun polymeric GN nonwoven mat was peeled off from the collector and then carbonized at 1000 °C for three hours under an inert atmosphere. This treatment converted the PAN fibers into CNFs and formed a 1D/2D CNF–GN–CNF heterojunction, as shown in Figure 4.22a–d. The GN/CNF-1, GN/CNF-2, and GN/CNF-3 composites contained 9.4, 17.2, and 31.9 wt% GN, respectively.

The transmission electron microscopy (TEM) images in Figure 4.22e–h and the corresponding schematics in Figure 4.22i–l demonstrate that two kinds of contacts are prevalent in the samples. In the case of neat CNFs, the point contact is the main contact mode between adjacent fibers, whereas in the presence of a CNF–GN–CNF heterojunction, 1D contact between GNs and CNFs can be achieved (Figure 4.22f,j). This type of heterojunction could increase the electrical conductivity. Similarly, the bridging mode between CNFs (Figure 4.22g,k) can also be modified by the introduction of GNs, with the CNF–GN–CNF heterojunction structures serving as bridges and conductive connections between CNFs to produce 2D conductive regions, as illustrated in Figure 4.22h,l. The benefit of such heterostructures on the conductivity can be seen in Figure 4.22m,n, with the electrical conductivity of the GN/CNF composites reaching 800 S m^{-1}. The high electrical conductivity originated from the reduced contact resistance between adjacent CNFs because of the insertion of GNs. However, in the case of high GN loading (31.9 wt%), the electrical conductivity was slightly decreased because CNFs with smaller diameters were produced, which also had a detrimental effect on the mechanical properties.

The EMI SE of a GN/CNF composite (17.2 wt%) and the neat CNF network are shown in Figure 4.22o. Clearly, the GN/CNF composite delivered a better EMI SE than the CNF network, which suggested that the CNF–GN–CNF heterojunction made a positive contribution to the shielding performance. The EMI SE of the CNF–GN–CNF heterojunction reached 25–28 dB at a thickness of 0.22–027 mm, which is sufficient to meet many commercial requirements. These values were 50–60% higher than those of the neat CNF network. Moreover, the SE_A and SE_R values of the CNF–GN–CNF heterojunction were also higher than those of the CNF network. It is important to note that the difference in EMI SE between the two samples was similar to the difference in electrical conductivity. Because the CNF–GN–CNF heterojunction structures were lightweight with a density of 0.08–0.1 g cm^{-3}, SSE/t was as high as 250 dB cm^2 g^{-1} at very small thicknesses of 0.22–0.27 mm.

4.3 Graphene as a Microwave Absorber

Traditionally, graphene has been used as an EMI shielding material owing to its excellent electrical conductivity. The EMI shielding mechanism generally depends

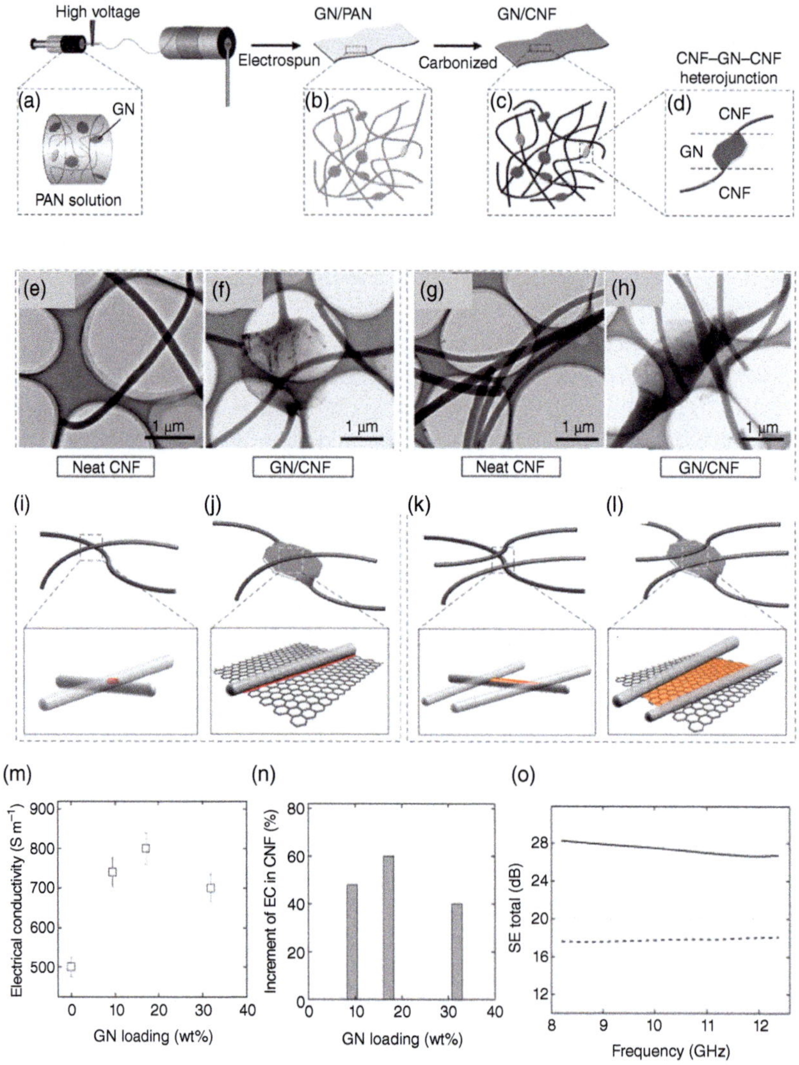

Figure 4.22 (a–d) Schematic of the fabrication process of GN/PAN and CNF–GN–CNF heterojunction structures. TEM images and corresponding schematics of (e, i) two adjacent CNFs in a neat CNF network, (f, j) a cross formed by a CNF over a CNF–GN–CNF heterojunction, (g, k) CNF bridges over adjacent CNFs, and (h, l) a bridge formed by a CNF–GN–CNF heterojunction. (m) Dependence of electrical conductivity on GN loading. (n) Increase in electrical conductivity (EC) in GN/CNF composite networks. (o) SE_T of a neat CNF network (dashed) and a GN/CNF composite network with 17.2 wt% GN (solid). Source: Song et al. [80]. Reproduced with permission of the American Chemical Society.

on two contributions, the first from surface reflection, which depends on the surface mobile charges, and the second from absorption inside the material. The reflection contribution depends heavily on the impedance mismatch between the material and the medium in which the EMWs are traveling. A larger difference in impedance between the two media enhances reflection from the surface. Therefore, it is beneficial to use materials with high electrical conductivity. In the case of absorption, the material does not need to be superconductive; instead, it should possess a moderate conductivity that is sufficient to create conductive channels. In contrast, when designing microwave absorbing materials, conductivity is not the primary tuning parameter because reflection has to be minimized. This type of application requires that all the incoming EMWs are absorbed inside the material without being reflected from the surface. EMI shielding materials that reflect EMWs can shield underlying equipment/components; however, the reflected waves are harmful to surrounding components. Therefore, to completely mitigate the influence of EMWs, microwave absorbers are designed with minimal impedance matching so that the incoming EMWs can penetrate the material. To achieve this, strong dielectric permittivity and permeability are required, which allow the material to absorb or dissipate the incoming EMWs in the form of heat. The microwave absorbing ability of a material is evaluated using the reflection loss (RL) or reflection coefficient (RC) and the attenuation constant (α). The experimental measurement method for measuring RL differs from that for determining the EMI SE of a material. As discussed in preceding Chapters 2 and 3, a two-port system is used to collect the S-parameters for EMI SE calculations. In contrast, RL is measured based on the assumption of zero transmission. Therefore, a highly conductive metal foil or plate with minimum transmission is used as a backing support to test the absorbing material. In this way, all the incident EMWs are reflected and/or absorbed without transmission. RL is expressed as follows [81]:

$$\text{RL} = 20\ \log\left|\frac{Z_{\text{in}} - Z_0}{Z_{\text{in}} + Z_0}\right| \tag{4.1}$$

Here, Z_{in} is the characteristic input impedance of the material and Z_0 is the impedance of air. Z_{in} is calculated as follows [81, 82]:

$$Z_{\text{in}} = \sqrt{\frac{\mu}{\varepsilon}}\ \tanh\left[j\left(\frac{2\pi fd}{c}\right)\sqrt{\mu\varepsilon}\right] \tag{4.2}$$

where ε and μ are the complex permittivity and the permeability of the absorbing material, respectively, and c is the speed of light.

Impedance matching at the interface between air and a lossy conducting shield ($Z_{\text{in}} \approx Z_0$) minimizes the reflection of EMWs from the surface. Impedance matching is the primary requirement for minimizing the RL value and also allows the maximum amount of EMWs to enter the shield. Once the EMWs enter the shield, a material with a larger attenuation constant exhibits maximum energy loss through absorption. As a low electrical conductivity is required for an absorber, the attenuation constant α can be calculated as follows [83, 84]:

$$\alpha = \frac{\sqrt{2}\pi f}{c} \times \sqrt{(\mu''\varepsilon'' - \mu'\varepsilon') + \sqrt{(\mu'\varepsilon'' + \mu''\varepsilon')^2 + (\mu''\varepsilon'' - \mu'\varepsilon')^2}} \tag{4.3}$$

Equation (4.3) shows that the attenuation constant is directly associated with the values of the complex dielectric permittivity ($\varepsilon = \varepsilon' - j\varepsilon''$) and the magnetic permeability ($\mu = \mu' - j\mu''$). The real parts of the complex permittivity (ε') and permeability (μ') indicate the capacitive energy storage of the electric and magnetic fields, whereas the imaginary parts (ε'') and (μ'') characterize all types of polarization losses in the form of heat dissipation. For a better understanding, the tangents of dielectric loss ($\tan\delta_\varepsilon = \frac{\varepsilon''}{\varepsilon'}$) and magnetic loss ($\tan\delta_\mu = \frac{\mu''}{\mu'}$) are used to show the loss mechanisms.

When the RL curves are plotted against frequency, there are sharp peak patterns. To understand the dependency on the frequency, a quarter-wavelength matching model needs to be considered. In fact, the peak frequency, also called the matching frequency (f_m), is related to the matching thickness (t_m) of the absorber and can be described as follows [85]:

$$t_m = \frac{n\lambda}{4} = \frac{nc}{(4f_m\sqrt{(|\varepsilon_r||\mu_r|)})} \tag{4.4}$$

where n is an integer ($n = 1, 2, 3, \ldots$), λ is the wavelength of the EMWs, and $|\varepsilon_r|$ and $|\mu_r|$ are the relative permittivity and permeability at f_m. At matching conditions of t_m and f_m, the EMWs at the air–absorber interface are out of phase with the absorber–metal interface by 180°. In this scenario, extinction of the EMWs will occur at the air–absorber interface, whereas maximum absorption will occur within the shield at f_m. An RL value of −20 dB is considered to attenuate 99% of the EMWs by the absorption mechanism. Efficient absorbing materials convert the absorbed energy of EMWs into thermal energy through magnetic loss and/or dielectric loss by balancing the relative permeability and/or permittivity.

The penetrating EMWs then interact with the host structure and are absorbed by different mechanisms such as eddy current loss, ohmic loss, polarization loss, and multiple internal reflections. Thus, the two mechanisms fundamentally differ from each other. Microwave absorbers tend to be less conductive but carry more magnetic constituents so that the absorption of EMWs can be realized. A reasonable electrical conductivity is required to provide a conductive medium. To produce high-performance electromagnetic wave absorbers (EMWAs), magnetic materials such as sendust, ferrites, and iron oxide have been used, with the conductivity supplemented by the addition of metallic fillers. However, metallic fillers are heavy and difficult to process. Thus, carbon-based fillers are being explored to not only reduce the weight and cost of EMWAs but also improve the processability to make the market competitive. In this direction, several researchers have used different carbon-based materials, including graphite, CNFs, CNTs, and graphene. Graphene provides advantages such as a large surface area, easy processability, easy fabrication, and dispersibility in various solvents. Moreover, if the intrinsic conductivity of the filler is high, only a minimal amount will be required to produce efficient EMWAs. Thus, graphene has been used in combination with many other nanomaterials to produce high-performance EMWAs. For example, Li et al. added Fe_3O_4 nanoparticles with graphene to improve the RL properties [86]. The Fe_3O_4–graphene hybrids were fabricated by the chemical thermolysis of iron ions attracted to GO through a simple polyol method. The Fe_3O_4–graphene hybrids were mixed with paraffin wax at 15 vol% for RL measurements. The magnetic properties of the Fe_3O_4–graphene hybrids and Fe_3O_4 nanoparticles, which were measured using a vibrating sample magnetometer, showed a typical S-like shape. Owing to the presence of graphene, the saturation magnetization of the Fe_3O_4–graphene hybrids was smaller than that of the Fe_3O_4 nanoparticles. Figure 4.23a shows a schematic of the interaction between EMWs and the Fe_3O_4–graphene hybrid material. The dissipation of incident EMWs is due to magnetic and dielectric losses. The magnetic loss originates from eddy current loss and the natural resonance of the Fe_3O_4 nanoparticles. The graphene sheets act as a medium that can disperse the electrostatic charge, resulting in high dielectric loss. The magnetic and dielectric losses are dissipated in the form of heat. The variation in RL as a function of thickness is shown in Figure 4.23b. No absorption peak was observed in the entire frequency region when the thickness of the filler material was 1 mm. The maximum RL value of −30.1 dB was observed at 17.2 GHz for a 1.48 mm thick sample. The RL results indicated that the Fe_3O_4–graphene hybrid composites with thicknesses of 1.48–3 mm have good absorption characteristics in the X- and Ku-bands. As observed in previous studies, the RL curve shifted toward lower frequencies as the thickness of the composite increased. The quarter-wavelength principal can be used to explain the wave absorption phenomenon. The relationship between the absorber thickness t_m and the peak frequency f_m can be described as the quarter-wavelength ($\lambda/4$) condition. Figure 4.23c shows a simulation of the absorber thickness vs. peak frequency for the Fe_3O_4–graphene hybrids. The simulations using the quarter-wavelength principle are in good agreement with the transmission line theory calculations for the Fe_3O_4–graphene hybrid composites. Figure 4.23d shows the frequency dependence of $Z = Z_{in}/Z_0$ for the Fe_3O_4–graphene hybrid composites (black dashed curve). The relationship between RL at the matching thickness and frequency is represented by the blue solid curve. For example, when the matching frequency was 17.2 GHz, the corresponding RL was the smallest with a value of −30 dB (blue solid curve, Figure 4.23d). Moreover, the relevant Z value (black dashed curve, Figure 4.23d) was close to 1 and the matching thickness was 1.48 mm (black solid curve, Figure 4.23c). This study showed that the quarter-wavelength principle can be used with the transmission line theory as a guide to design EMW materials for specific applications.

Feng et al. utilized the interfacial interactions and synergistic effect of CoNi nanocrystals and nitrogen-doped graphene (NG) in a composite for microwave absorption [87]. The CoNi/NG hybrid was prepared by a one-step polyol process. Ni^{2+} and Co^{2+} were first added to a GO and ethylene glycol (EG) suspension to coordinate with the oxygen-containing functional groups on GO. Then, $N_2H_4{\cdot}H_2O$ containing NaOH was added to the solution and heated at 120 °C. During this process, CoNi nanocrystals were nucleated and grew on the surfaces of NG nanosheets, leading to the formation of the CoNi/NG hybrid, as shown in Figure 4.24a. The CoNi/NG sample exhibited less saturation magnetization than CoNi mainly because of the presence of nonmagnetic NG (Figure 4.24b). The microwave absorption performance is largely related to the relative complex permittivity and permeability, in which the real parts (ε' and μ') and imaginary parts (ε'' and μ'') represent the storage capability and dissipation capabilities for both the electrical and magnetic energy. Figure 4.24c,d shows the complex permittivity and permeability of GO, CoNi nanocrystals, and CoNi/NG hybrids. Owing to the absence of a magnetic

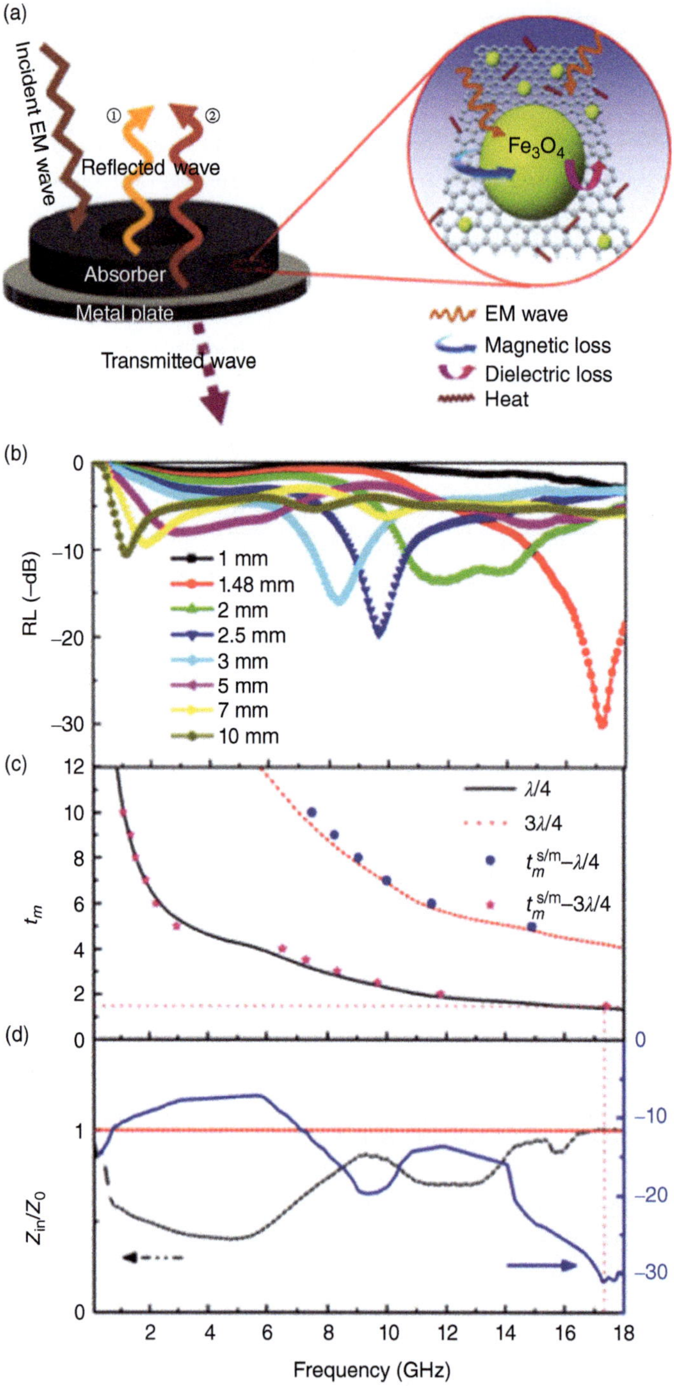

Figure 4.23 (a) Schematic representation of the interactions of EMWs with a microwave absorber. (b) RL values vs. frequency for Fe_3O_4-graphene/paraffin hybrids with various thicknesses. (c) Simulations of the absorber thickness t_m vs. peak frequency f_m for Fe_3O_4-graphene/paraffin. (d) Modulus of the normalized input impedance $|Z_{in}/Z_0|$ generated from the Fe_3O_4-graphene hybrids. Source: Li et al. [86]. Reproduced with permission of Springer.

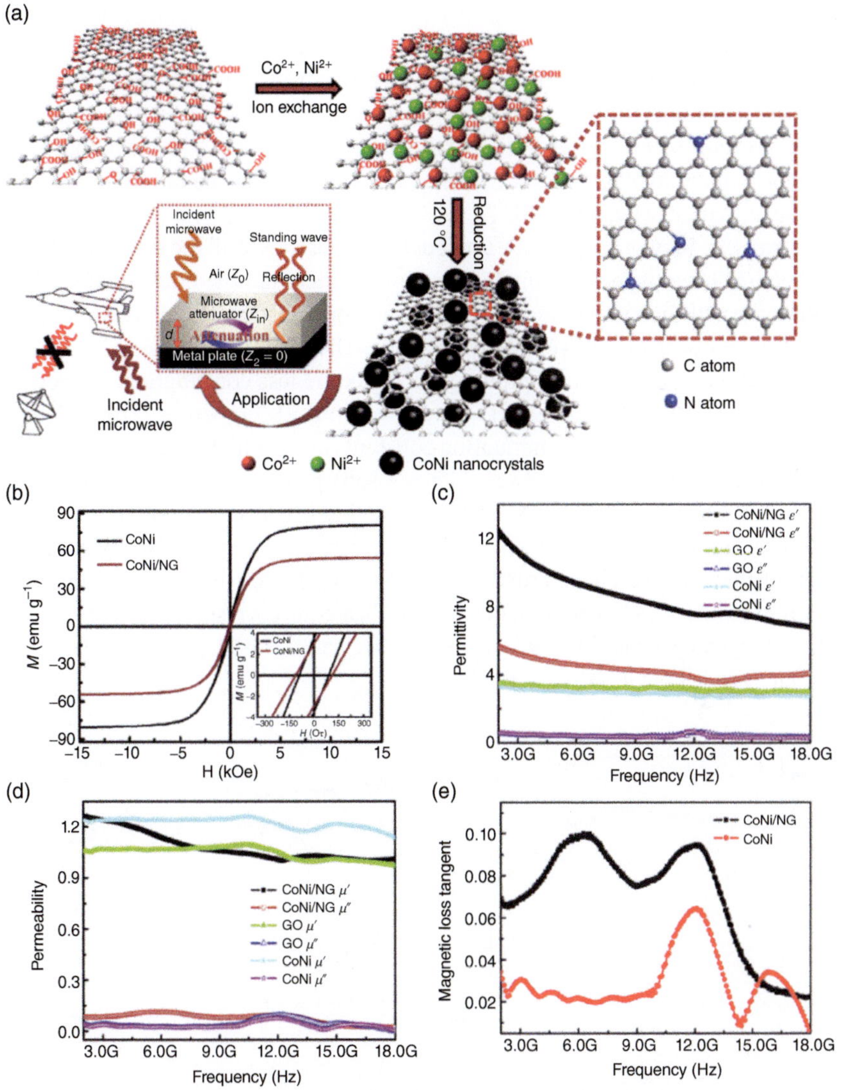

Figure 4.24 (a) Schematic of the formation mechanism of CoNi/NG hybrids. (b) Magnetic hysteresis loops of CoNi nanocrystals and CoNi/NG hybrids. (c) Complex permittivities and (d) complex permeabilities of CoNi nanocrystals, GO, and CoNi/NG hybrids. (e) Magnetic loss tangents of CoNi nanocrystals and CoNi/NG hybrids. Source: Feng et al. [87]. Reproduced with permission of Elsevier.

constituent, GO exhibited the lowest complex permeability, whereas the complex permeabilities of the other samples were reasonable. CoNi exhibited a low complex permittivity; in contrast, the high complex permittivity of the CoNi/NG sample was ascribed to the relaxation and polarization of the residual defects and groups on the NG surface, which can act as polarization centers to promote electric dipole polarization, interfacial polarization, and relaxation polarization. The magnetic and dielectric loss tangents are effective parameters for investigating the magnetic and dielectric loss capacities of a material. Figure 4.24e shows the magnetic loss tangents of CoNi and the CoNi/NG hybrids. The introduction of graphene increased the magnetic loss tangent of CoNi/NG, which could be due to the NG sheets, preventing the agglomeration of the CoNi nanocrystals. The magnetic loss in the hybrid material originated from the eddy current effect, the natural resonance, and the exchange resonance [88].

Figure 4.25a shows the dielectric tangent loss properties of all the samples. Clearly, the CoNi/NG sample exhibited a higher tangent loss than the other samples. This result indicated that graphene and CoNi have a synergistic effect on improving the dielectric loss over the entire measured range. The residual groups and defects in NG played a crucial role in improving the dielectric loss contribution. Microwave attenuation is also a key parameter when studying RL properties. The CoNi/NG hybrid exhibited the highest attenuation constant (m^{-1}), which was particularly significant in the high-frequency range (Figure 4.25b), providing direct evidence for the absorption properties of the CoNi/NG hybrid being superior to those of the pristine materials. Figure 4.25c shows the RL curves vs. frequency for CoNi/NG hybrids with different thicknesses. The simulation of t_m (d_m) vs. f_m for the CoNi/NG hybrid is shown in Figure 4.25d, where the red stars indicate the corresponding thickness vs. critical frequency obtained directly from the RL curves. The RL peak shifted to lower frequencies with the increase in thickness.

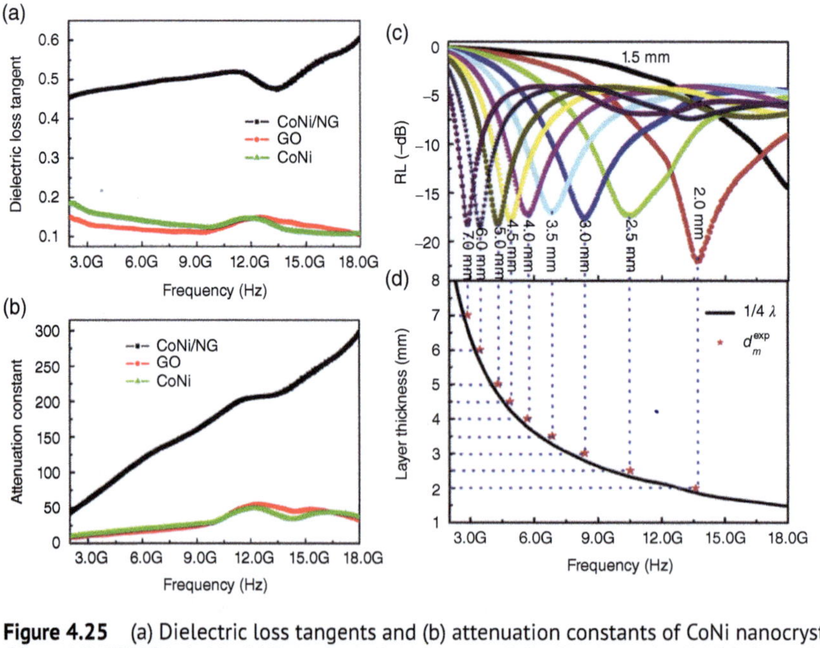

Figure 4.25 (a) Dielectric loss tangents and (b) attenuation constants of CoNi nanocrystals, GO, and CoNi/NG hybrids. (c) Frequency dependence of the RL curves for CoNi/NG hybrids with various thicknesses. (d) Simulations of the absorber thickness vs. peak frequency (f_m) for the CoNi/NG hybrids using the quarter-wavelength model. Source: Feng et al. [87] Reproduced with permission of Elsevier.

In another approach, Quan et al. improved the impedance-matching conditions by transforming graphene from diamagnetic to ferromagnetic while suppressing its conductivity [89]. Graphene was doped with nitrogen via a hydrothermal approach using urea as a nitrogen precursor (Figure 4.26a). The NG samples exhibited magnetism via a Ruderman–Kittel–Kasuya–Yosida (RKKY) mechanism [90]. Among the three types of nitrogen-doped structures, pyrrolic-N exhibited magnetic properties, which in combination with a reduced electrical conductivity was found to be more suitable for microwave absorption applications. The samples were prepared in paraffin wax at 30 wt% loading. The NG samples had a maximum nitrogen doping content of 5 at.%. The GO sample exhibited a very weak paramagnetic response, whereas RGO showed higher magnetization and exhibited ferromagnetic behavior owing to the removal of oxygen-containing groups, induced vacancies, and the distortion of the graphene lattice, which led to a magnetic moment (Figure 4.26b) [91]. The ferromagnetic behavior became more pronounced upon nitrogen doping, with NG showing six times higher magnetization (emu g^{-1}) than GO. Zero field cooling (ZFC) and field cooling (FC) measurements were also performed in the temperature range of 5–300 K. The ZFC and FC curves for GO, RGO, and NG showed superparamagnetic behavior, where the separation between the two curves increased below the blocking temperature (T_B) with decreasing temperature until 5 K. The T_B values followed the order GO > RGO > NG, which indicated the transformation from magnetically disordered to ordered systems. This result implied that NG requires the least reverse energy among the three samples. T_B refers to the threshold point of a thermally activated system. Above this temperature, magnetic order may still exist; however, each particle behaves such that it is in a paramagnetic state. In contrast, below T_B, the magnetization direction of each nanoparticle aligns with its easy axis [92]. The interaction between the localized electron spin constituted as a result of GO reduction; that is, the RKKY interaction provides an additional anisotropy barrier, and thus, less thermal energy was required for the magnetic phase transformation [93]. In the NG samples, the nitrogen dopants reduced the distance between the magnetic moments, and hence, the RKKY magnetic interactions were greatly enhanced. However, it is important to note that a significant ferromagnetic response, even from NG, could not be achieved at room temperature.

All the samples subjected to RL measurements showed a typical loss curve, with the NG sample exhibiting the strongest response in the 12–14 GHz range (Figure 4.26c). The improved magnetic and electrical properties of NG resulted in better impedance matching. The additional vacancy defects caused by nitrogen doping also created new polarization centers that aided in EMW absorption. The residual oxygenated groups, resulting from insufficient reduction, also helped to create electric dipole polarization. It is of utmost importance to design materials for RL that have small thicknesses. Because thickness plays a crucial role in shifting the maximum RL in the frequency spectra, NG samples with different thicknesses were studied (Figure 4.26d). To better understand the microwave absorption behavior, the quarter-wavelength ($\lambda/4$) principle was adopted. The matching thickness and the corresponding peak frequency were calculated. As shown in Figure 4.26e,

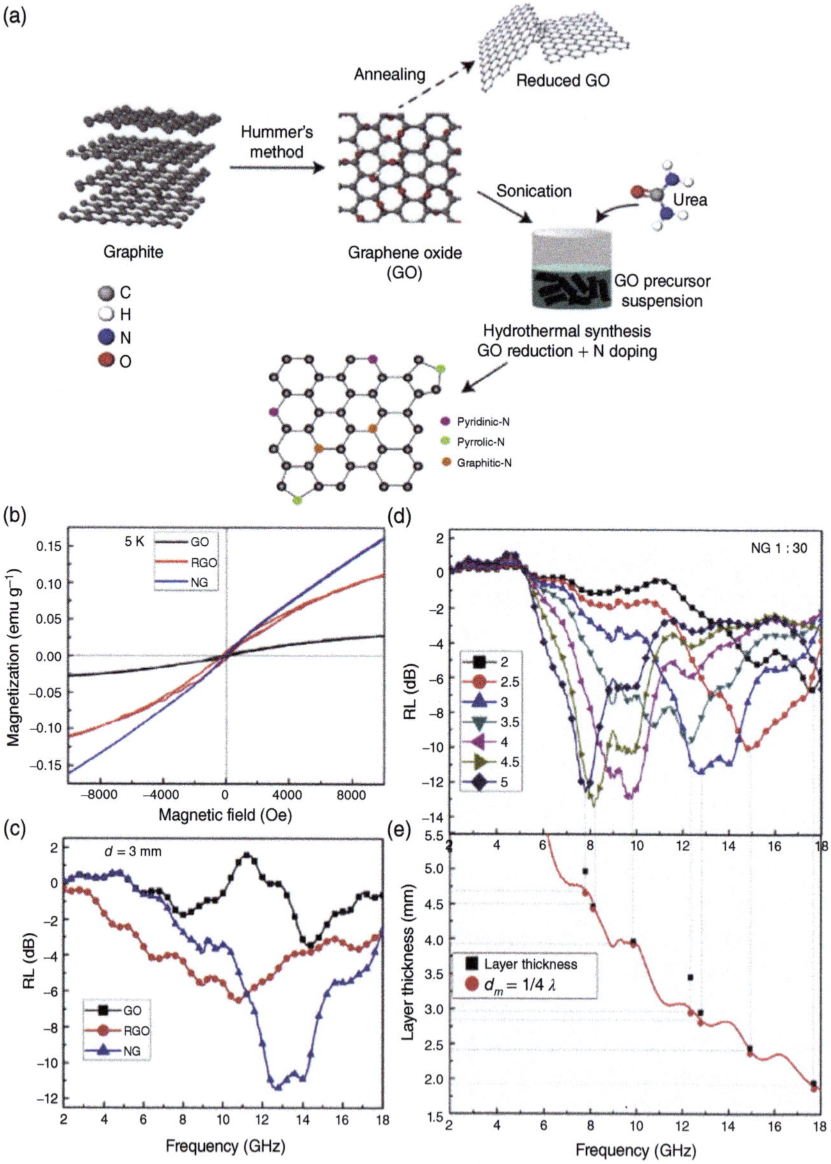

Figure 4.26 (a) Schematic of the synthesis processes of GO, RGO, and NG. (b) Hysteresis loops of GO, RGO, and NG30 at 5 K. (c) Reflection losses of GO, RGO, and NG. (d) Dependence of the RL curves of NG on frequency at various thicknesses. (e) Dependence of the thickness on frequency for NG powder/paraffin, as calculated using the $\lambda/4$ model. The dashed lines in (d) indicate the thickness contours at 1.91, 2.41, 2.84, 2.99, 3.94, 4.46, and 4.68 mm. Source: Quan et al. [89]. Reproduced with permission of Elsevier.

the matching thickness (red dots) at the peak frequency lies very close to the experimental thicknesses (black dots), thus satisfying the $\lambda/4$ model.

In another interesting approach, Zhu et al. developed flower-like $CoS_2@MoS_2$ core–shell microspheres and coated them with RGO for an effective RL demonstration [94]. First, the CoS_2 nanospheres were prepared from $CoCl_2{\cdot}H_2O$ and sulfur powder with the addition of polyvinylpyrrolidone (PVP). Hydrothermal treatment at 200 °C for 12 hours yielded a black precipitate of CoS_2, which was subsequently dispersed in cetyltrimethylammonium bromide (CTAB). Then, $Na_2MoO_4{\cdot}2H_2O$ and L-cysteine were added slowly to the CoS_2/CTAB solution (Figure 4.27a), and the mixture was subjected to hydrothermal treatment at 200 °C for 12 hours. The $CoS_2@MoS_2$/rGO composites were prepared by mixing a $CoS_2@MoS_2$/water dispersion with a GO/water dispersion and heating in an autoclave at 200 °C for 24 hours. The core–shell powders were mixed with paraffin to form polymer composites S1 (10 wt% filler), S2 (20 wt% filler), and S3 (30 wt%

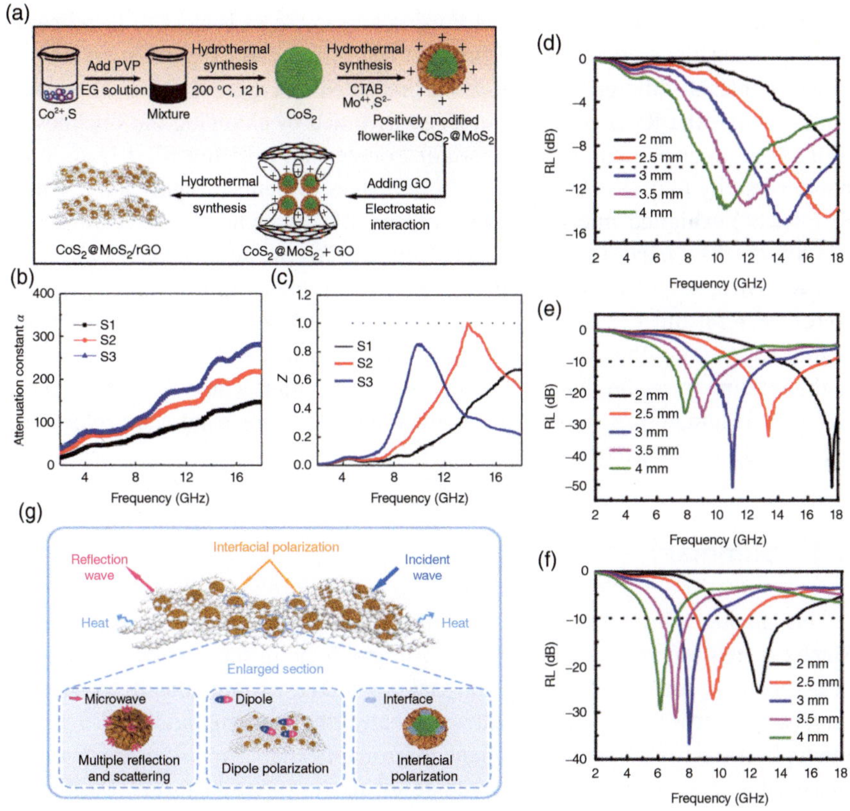

Figure 4.27 (a) Schematic of the fabrication of $CoS_2@MoS_2$/rGO. (b) Attenuation constant (α) and (c) impedance (Z) of the as-synthesized composites (at 2.4 mm). Reflection losses of samples (d) S1, (e) S2, and (f) S3 at different thicknesses. (g) Schematic of the absorption mechanisms of the $CoS_2@MoS_2$/rGO composite. Source: Zhu et al. [94]. Reproduced with permission of Elsevier.

filler) for RL measurements. The introduction of RGO on flower-like CoS_2@MoS_2 greatly improved the permittivity of the composite samples. The high values of the real part of the dielectric permittivity were related to enhanced dipolar polarization and interfacial polarization owing to the abundant residual defects from the synthesis process and the large interfacial area between CoS_2, MoS_2, and RGO. The enhancement in conductivity due to the presence of RGO also promoted dielectric loss. The Cole–Cole plot of the composite revealed many small circles, indicating the occurrence of complex multi-relaxation dielectric behavior originating from the unique flower-like core–shell structure of CoS_2@MoS_2 and the defect polarization owing to the oxygenated groups on RGO.

The attenuation constant, which can be used to evaluate the general absorption characteristics of a material, is presented in Figure 4.27b for all the samples. The samples with higher filler loadings exhibited better attenuation constants. The high attenuation constants were the result of the unique microstructure and the presence of multiple scattering points, which promoted the dissipation of microwave energy in the form of heat. The calculated impedance (Z) values are also presented in Figure 4.27c, where the Z value for sample S2 is close to 1, indicating perfect impedance-matching conditions. Furthermore, the RL performance of the three samples at different thicknesses is presented in Figure 4.27d–f. The results indicated that the RL properties can be easily tuned by changing the thickness as well as the filler content. It is important to note that a higher filler loading did not necessarily improve the RL contribution, as is evident in Figure 4.27. Sample S3 (30 wt%) exhibited inferior impedance and RL values as compared to sample S2 (20 wt%). The maximum RL values of samples S1, S2, and S3 were − 15.1, −58, and − 52.3 dB, respectively. Moreover, the effective bandwidths of these samples were 4.96, 6.24, and 4.72 GHz at thicknesses of 2.90, 2.40, and 1.70 mm, respectively. Figure 4.27g shows the possible absorption mechanisms, with multiple reflections and scattering from the flower-like microstructure, interfacial polarization due to different nanomaterials, and dipole polarization, all playing significant roles in the excellent RL performance.

4.4 Summary

In this chapter, a brief overview of the different forms of graphene was presented. Since the first report on EMI shielding with graphene in the past decade, the use of graphene as an EMI shielding material has increased dramatically because its intrinsic properties are superior to those of many other nanomaterials. Also, having the highest electron mobility, graphene is strong, flexible, corrosion-resistant, and processable, making it an ideal candidate for use in different forms. Despite the outstanding progress with graphene in the EMI shielding field, several challenges still need to be addressed. The reduced form of GO is largely hydrophobic, which is challenging for solution dispersion and preparing polymer composites, and the agglomeration of RGO is a common issue in some polymers. In various reports, large-area graphene sheets have been used to improve the electrical conductivity and EMI SE; however, such large-area sheets are more difficult to exfoliate than small-area graphene sheets. Furthermore, the chemical methods used to synthesize graphene impart several defects that need to be controlled, either via modification of the synthesis process or heteroatom doping. Similarly, the contact resistance between the graphene sheets can be significantly reduced by using large graphene sheets. These strategies generally aid in improving the intrinsic properties of graphene or its derivate RGO for subsequent processing steps.

For many shielding applications, graphene must be embedded in polymer matrices to construct mechanically strong composite materials. Although PG can deliver a high shielding efficiency, it is not always cost-effective to utilize pure graphene for every application. Instead, the addition of a low filler content in the range of 1–20 wt% can render polymer composites acceptable for many shielding applications. This approach not only minimizes the amount of required graphene but also produces materials that are mechanically robust and can sustain harsh environments. Recently, hydrophobic composites have been synthesized, including wearable jackets, for outdoor applications. However, it is important that graphene is properly dispersed in the polymer without any agglomeration. The dispersion ability of graphene can be further improved by surface functionalization. Common polymers such as polyvinyl alcohol (PVA), polypropylene (PP), polystyrene (PS), polyvinyl chloride (PVC), and polyvinylidene fluoride (PVDF) have shown excellent compatibility with graphene and provide a good balance between the shielding efficiency and the mechanical properties.

The properties of graphene can be further enhanced by doping graphene with heteroatoms such as nitrogen, sulfur, boron, phosphorus, and chlorine to heal the majority of defects. These heteroatoms impart either n-type or p-type behavior. In particular, n-type dopants are favorable because they provide extra electrons to the graphene lattice, which improves the electron density and thus the electrical conductivity. Although there have been numerous reports on doped graphene for shielding applications, codoping studies are rare. Thus, there is considerable potential for exploring codopants in the graphene lattice for shielding. Codopants may provide extra polarization centers in addition to improving the electrical properties. However, it is also crucial to develop a doping method that is scalable, repeatable, easy to manage, and cost-effective. In particular, the search for biomass-derived dopants as inexpensive alternatives should be accelerated.

The shielding properties of graphene can be further enhanced by forming hybrids with other carbon-based materials. In such hybrids, the shielding efficiency is not only improved via the conductivity contribution but also by different types of polarization at the interface. 1D carbon nanofillers such as CNFs and CNTs can act as bridges between graphene sheets, thus increasing the conduction. Point and bridge contacts create extra polarization centers that are helpful in mitigating EMI. It is also important to develop carbon-based materials from biomass through suitable processing steps. Some examples, such as sugarcane and rice husks, have already been reported to provide conductive materials after high-temperature treatments. This approach will not only provide low-cost graphene/C structures but can also utilize the huge amounts of biomass waste that contribute to environmental pollution.

References

1 Ferrari, A.C., Meyer, J., Scardaci, V. et al. (2006). Raman spectrum of graphene and graphene layers. *Physical Review Letters* 97 (18): 187401.

2 Castro Neto, A.H., Guinea, F., Peres, N.M.R. et al. (2009). The electronic properties of graphene. *Reviews of Modern Physics* 81 (1): 109–162.

3 Nilsson, J., Neto, A.C., Guinea, F., and Peres, N. (2006). Electronic properties of graphene multilayers. *Physical Review Letters* 97 (26): 266801.

4 Pulizzi, F., Bubnova, O., Milana, S. et al. (2019). Graphene in the making. *Nature Nanotechnology* 14 (10): 914–918.

5 Chua, C.K. and Pumera, M. (2013). Covalent chemistry on graphene. *Chemical Society Reviews* 42 (8): 3222–3233.

6 Novoselov, K.S., Geim, A.K., Morozov, S.V. et al. (2004). Electric field effect in atomically thin carbon films. *Science* 306 (5696): 666–669.

7 Novoselov, K.S., Geim, A.K., Morozov, S. et al. (2005). Two-dimensional gas of massless Dirac fermions in graphene. *Nature* 438 (7065): 197–200.

8 Ohta, T., Bostwick, A., Seyller, T. et al. (2006). Controlling the electronic structure of bilayer graphene. *Science* 313 (5789): 951–954.

9 Liu, H., Liu, Y., and Zhu, D. (2011). Chemical doping of graphene. *Journal of Materials Chemistry* 21 (10): 3335–3345.

10 Farmer, D.B., Golizadeh-Mojarad, R., Perebeinos, V. et al. (2009). Chemical doping and electron – hole conduction asymmetry in graphene devices. *Nano Letters* 9 (1): 388–392.

11 Novoselov, K.S., Jiang, Z., Zhang, Y. et al. (2007). Room-temperature quantum Hall effect in graphene. *Science* 315 (5817): 1379–1379.

12 Geim, A.K. and Novoselov, K.S. (2010). The rise of graphene. In: *Nanoscience and Technology: A Collection of Reviews from Nature Journals* (ed. P. Rodgers), 11–19. World Scientific.

13 Du, X., Skachko, I., Barker, A., and Andrei, E.Y. (2008). Approaching ballistic transport in suspended graphene. *Nature Nanotechnology* 3 (8): 491-495.

14 Malekpour, H., Chang, K.-H., Chen, J.-C. et al. (2014). Thermal conductivity of graphene laminate. *Nano Letters* 14 (9): 5155–5161.

15 Lee, C., Wei, X., Kysar, J.W., and Hone, J. (2008). Measurement of the elastic properties and intrinsic strength of monolayer graphene. *Science* 321 (5887): 385–388.

16 Sheehy, D.E. and Schmalian, J. (2009). Optical transparency of graphene as determined by the fine-structure constant. *Physical Review B* 80 (19): 193411.

17 Economopoulos, S.P. and Tagmatarchis, N. (2013). Chemical functionalization of exfoliated graphene. *Chemistry – A European Journal* 19 (39): 12930–12936.

18 Kuila, T., Bose, S., Mishra, A.K. et al. (2012). Chemical functionalization of graphene and its applications. *Progress in Materials Science* 57 (7): 1061–1105.

19 Boukhvalov, D. and Katsnelson, M. (2009). Chemical functionalization of graphene. *Journal of Physics: Condensed Matter* 21 (34): 344205.

20 Yeh, T.-F., Cihlář, J., Chang, C.-Y. et al. (2013). Roles of graphene oxide in photocatalytic water splitting. *Materials Today* 16 (3): 78–84.

21 Sathe, B.R., Zou, X., and Asefa, T. (2014). Metal-free B-doped graphene with efficient electrocatalytic activity for hydrogen evolution reaction. *Catalysis Science & Technology* 4 (7): 2023–2030.

22 Chai, G.-L., Qiu, K., Qiao, M. et al. (2017). Active sites engineering leads to exceptional ORR and OER bifunctionality in P,N Co-doped graphene frameworks. *Energy & Environmental Science* 10 (5): 1186–1195.

23 Zurutuza, A. and Marinelli, C. (2014). Challenges and opportunities in graphene commercialization. *Nature Nanotechnology* 9 (10): 730–734.

24 Lin, L., Peng, H., and Liu, Z. (2019). Synthesis challenges for graphene industry. *Nature Materials* 18 (6): 520–524.

25 Novoselov, K.S., Fal, V., Colombo, L. et al. (2012). A roadmap for graphene. *Nature* 490 (7419): 192–200.

26 Thomassin, J.-M., Jerome, C., Pardoen, T. et al. (2013). Polymer/carbon based composites as electromagnetic interference (EMI) shielding materials. *Materials Science & Engineering R: Reports* 74 (7): 211–232.

27 Du, J. and Cheng, H.M. (2012). The fabrication, properties, and uses of graphene/polymer composites. *Macromolecular Chemistry and Physics* 213 (10–11): 1060–1077.

28 Kim, H., Abdala, A.A., and Macosko, C.W. (2010). Graphene/polymer nanocomposites. *Macromolecules* 43 (16): 6515–6530.

29 Jia, L.-C., Yan, D.-X., Liu, X. et al. (2018). Highly efficient and reliable transparent electromagnetic interference shielding film. *ACS Applied Materials and Interfaces* 10 (14): 11941–11949.

30 Erdogan, N., Erden, F., Astarlioglu, A.T. et al. (2020). ITO/Au/ITO multilayer thin films on transparent polycarbonate with enhanced EMI shielding properties. *Current Applied Physics* 20 (4): 489–497.

31 Wang, H., Lu, Z., Liu, Y. et al. (2017). Double-layer interlaced nested multi-ring array metallic mesh for high-performance transparent electromagnetic interference shielding. *Optics Letters* 42 (8): 1620–1623.

32 Huang, J.-L., Yau, B.-S., Chen, C.-Y. et al. (2001). The electromagnetic shielding effectiveness of indium tin oxide films. *Ceramics International* 27 (3): 363–365.

33 Hong, S.K., Kim, K.Y., Kim, T.Y. et al. (2012). Electromagnetic interference shielding effectiveness of monolayer graphene. *Nanotechnology* 23 (45): 455704.

34 Chen, J.-H., Jang, C., Xiao, S. et al. (2008). Intrinsic and extrinsic performance limits of graphene devices on SiO_2. *Nature Nanotechnology* 3 (4): 206–209.

35 Zhang, N., Wang, Z., Song, R. et al. (2019). Flexible and transparent graphene/silver-nanowires composite film for high electromagnetic interference shielding effectiveness. *Science Bulletin* 64 (8): 540–546.

36 Han, Y., Liu, Y., Han, L. et al. (2017). High-performance hierarchical graphene/metal-mesh film for optically transparent electromagnetic interference shielding. *Carbon* 115: 34–42.

37 Ma, L., Lu, Z., Tan, J. et al. (2017). Transparent conducting graphene hybrid films to improve electromagnetic interference (EMI) shielding performance of graphene. *ACS Applied Materials and Interfaces* 9 (39): 34221–34229.

38 Kim, S., Oh, J.-S., Kim, M.-G. et al. (2014). Electromagnetic interference (EMI) transparent shielding of reduced graphene oxide (RGO) interleaved structure fabricated by electrophoretic deposition. *ACS Applied Materials and Interfaces* 6 (20): 17647–17653.

39 Zhang, H.-L., Xia, Y., and Gai, J.-G. (2018). Ultrathin active layer for transparent electromagnetic shielding window. *ACS Omega* 3 (3): 2765–2772.

40 Lu, Z., Ma, L., Tan, J. et al. (2016). Transparent multi-layer graphene/polyethylene terephthalate structures with excellent microwave absorption and electromagnetic interference shielding performance. *Nanoscale* 8 (37): 16684–16693.

41 Kumar, P., Shahzad, F., Yu, S. et al. (2015). Large-area reduced graphene oxide thin film with excellent thermal conductivity and electromagnetic interference shielding effectiveness. *Carbon* 94: 494–500.

42 Shen, B., Zhai, W., and Zheng, W. (2014). Ultrathin flexible graphene film: an excellent thermal conducting material with efficient EMI shielding. *Advanced Functional Materials* 24 (28): 4542–4548.

43 Shim, J., Yun, J.M., Yun, T. et al. (2014). Two-minute assembly of pristine large-area graphene based films. *Nano Letters* 14 (3): 1388–1393.

44 Pham, V.H., Cuong, T.V., Hur, S.H. et al. (2010). Fast and simple fabrication of a large transparent chemically-converted graphene film by spray-coating. *Carbon* 48 (7): 1945–1951.

45 Xin, G., Zhu, W., Yao, T. et al. (2017). Microstructure control of macroscopic graphene paper by electrospray deposition and its effect on thermal and electrical conductivities. *Applied Physics Letters* 110 (9): 091909.

46 Zhao, J., Pei, S., Ren, W. et al. (2010). Efficient preparation of large-area graphene oxide sheets for transparent conductive films. *ACS Nano* 4 (9): 5245–5252.

47 Polsen, E.S., McNerny, D.Q., Viswanath, B. et al. (2015). High-speed roll-to-roll manufacturing of graphene using a concentric tube CVD reactor. *Scientific Reports* 5 (1): 1–12.

48 Wei, Q., Pei, S., Qian, X. et al. (2020). Superhigh electromagnetic interference shielding of ultrathin aligned pristine graphene nanosheets film. *Advanced Materials* 32 (14): 1907411.

49 Wilson, R., George, G., and Joseph, K. (2020). An introduction to materials for potential EMI shielding applications: status and future. In: *Materials for Potential EMI Shielding Applications* (eds. J. Kuruvilla, R. Wilson and G. Gejo), 1–8. Elsevier.

50 Abbasi, H., Antunes, M., and Velasco, J.I. (2019). Recent advances in carbon-based polymer nanocomposites for electromagnetic interference shielding. *Progress in Materials Science* 103: 319–373.

51 Shahzad, F., Lee, S.H., Hong, S.M., and Koo, C.M. (2018). Segregated reduced graphene oxide polymer composite as a high performance electromagnetic interference shield. *Research on Chemical Intermediates* 44 (8): 4707–4719.

52 Song, W.-L., Guan, X.-T., Fan, L.-Z. et al. (2015). Magnetic and conductive graphene papers toward thin layers of effective electromagnetic shielding. *Journal of Materials Chemistry A* 3 (5): 2097–2107.

53 Kumar, P., Narayan Maiti, U., Sikdar, A. et al. (2019). Recent advances in polymer and polymer composites for electromagnetic interference shielding: review and future prospects. *Polymer Reviews* 59 (4): 687–738.

54 Koo, C.M., Shahzad, F., Kumar, P. et al. (2018). Polymer-based EMI shielding materials. In: *Advanced Materials for Electromagnetic Shielding: Fundamentals, Properties, and Applications* (eds. M. Jaroszewski, S. Thomas and A.V. Rane), 177–217. Wiley.

55 Liang, J., Wang, Y., Huang, Y. et al. (2009). Electromagnetic interference shielding of graphene/epoxy composites. *Carbon* 47 (3): 922–925.

56 Dalal, J., Lather, S., Gupta, A. et al. (2018). EMI shielding properties of laminated graphene and $PbTiO_3$ reinforced poly (3,4-ethylenedioxythiophene) nanocomposites. *Composites Science and Technology* 165: 222–230.

57 Hu, H., Zhang, G., Xiao, L. et al. (2012). Preparation and electrical conductivity of graphene/ultrahigh molecular weight polyethylene composites with a segregated structure. *Carbon* 50 (12): 4596–4599.

58 Pang, H., Chen, T., Zhang, G. et al. (2010). An electrically conducting polymer/graphene composite with a very low percolation threshold. *Materials Letters* 64 (20): 2226–2229.

59 Ding-Xiang, Y., Huan, P., Ling, X. et al. (2014). Electromagnetic interference shielding of segregated polymer composite with an ultralow loading of in situ thermally reduced graphene oxide. *Nanotechnology* 25 (14): 145705.

60 Yan, D.X., Pang, H., Li, B. et al. (2015). Structured reduced graphene oxide/polymer composites for ultra-efficient electromagnetic interference shielding. *Advanced Functional Materials* 25 (4): 559–566.

61 Jiang, X., Yan, D.-X., Bao, Y. et al. (2015). Facile, green and affordable strategy for structuring natural graphite/polymer composite with efficient electromagnetic interference shielding. *RSC Advances* 5 (29): 22587–22592.

62 Sachdev, V.K., Patel, K., Bhattacharya, S., and Tandon, R.P. (2011). Electromagnetic interference shielding of graphite/acrylonitrile butadiene styrene composites. *Journal of Applied Polymer Science* 120 (2): 1100–1105.

63 Zhang, H.-B., Yan, Q., Zheng, W.-G. et al. (2011). Tough graphene– polymer microcellular foams for electromagnetic interference shielding. *ACS Aapplied Materials and Interfaces* 3 (3): 918–924.

64 Shen, B., Li, Y., Zhai, W., and Zheng, W. (2016). Compressible graphene-coated polymer foams with ultralow density for adjustable electromagnetic interference (EMI) shielding. *ACS Applied Materials and Interfaces* 8 (12): 8050–8057.

65 Wu, Y., Wang, Z., Liu, X. et al. (2017). Ultralight graphene foam/conductive polymer composites for exceptional electromagnetic interference shielding. *ACS Applied Materials and Interfaces* 9 (10): 9059–9069.

66 Shahzad, F., Kumar, P., Yu, S. et al. (2015). Sulfur-doped graphene laminates for EMI shielding applications. *Journal of Materials Chemistry C* 3 (38): 9802–9810.

67 Shahzad, F., Kumar, P., Kim, Y.-H. et al. (2016). Biomass-derived thermally annealed interconnected sulfur-doped graphene as a shield against electromagnetic interference. *ACS Applied Materials and Interfaces* 8 (14): 9361–9369.
68 Umrao, S., Gupta, T.K., Kumar, S. et al. (2015). Microwave-assisted synthesis of boron and nitrogen co-doped reduced graphene oxide for the protection of electromagnetic radiation in Ku-band. *ACS Applied Materials and Interfaces* 7 (35): 19831–19842.
69 Yang, Q.-H., Hou, P.-X., Unno, M. et al. (2005). Dual Raman features of double coaxial carbon nanotubes with N-doped and B-doped multiwalls. *Nano Letters* 5 (12): 2465–2469.
70 Sun, C.-L., Wang, H.-W., Hayashi, M. et al. (2006). Atomic-scale deformation in N-doped carbon nanotubes. *Journal of the American Chemical Society* 128 (26): 8368–8369.
71 Chhetri, S., Adak, N.C., Samanta, P. et al. (2019). Synergistic effect of Fe_3O_4 anchored N-doped rGO hybrid on mechanical, thermal and electromagnetic shielding properties of epoxy composites. *Composites Part B: Engineering* 166: 371–381.
72 Wan, Y.-J., Zhu, P.-L., Yu, S.-H. et al. (2017). Graphene paper for exceptional EMI shielding performance using large-sized graphene oxide sheets and doping strategy. *Carbon* 122: 74–81.
73 Wang, X.-X., Ma, T., Shu, J.-C., and Cao, M.-S. (2018). Confinedly tailoring Fe_3O_4 clusters-NG to tune electromagnetic parameters and microwave absorption with broadened bandwidth. *Chemical Engineering Journal* 332: 321–330.
74 Liu, Y., Xu, Z., Zhan, J. et al. (2016). Superb electrically conductive graphene fibers via doping strategy. *Advanced Materials* 28 (36): 7941–7947.
75 Wu, Z., Han, Y., Huang, R. et al. (2014). Semimetallic-to-metallic transition and mobility enhancement enabled by reversible iodine doping of graphene. *Nanoscale* 6 (21): 13196–13202.
76 Verma, M., Chauhan, S.S., Dhawan, S., and Choudhary, V. (2017). Graphene nanoplatelets/carbon nanotubes/polyurethane composites as efficient shield against electromagnetic polluting radiations. *Composites Part B: Engineering* 120: 118–127.
77 Singh, K., Ohlan, A., Pham, V.H. et al. (2013). Nanostructured graphene/Fe_3O_4 incorporated polyaniline as a high performance shield against electromagnetic pollution. *Nanoscale* 5 (6): 2411–2420.
78 Kong, L., Yin, X., Yuan, X. et al. (2014). Electromagnetic wave absorption properties of graphene modified with carbon nanotube/poly(dimethyl siloxane) composites. *Carbon* 73: 185–193.
79 Huangfu, Y., Ruan, K., Qiu, H. et al. (2019). Fabrication and investigation on the PANI/MWCNT/thermally annealed graphene aerogel/epoxy electromagnetic interference shielding nanocomposites. *Composites Part A: Applied Science and Manufacturing* 121: 265–272.
80 Song, W.-L., Wang, J., Fan, L.-Z. et al. (2014). Interfacial engineering of carbon nanofiber–graphene–carbon nanofiber heterojunctions in flexible lightweight

electromagnetic shielding networks. *ACS Applied Materials and Interfaces* 6 (13): 10516–10523.

81 Kaiser, K.L. (2005). *Electromagnetic Shielding*. CRC Press.

82 Ott, H.W. (2011). *Electromagnetic Compatibility Engineering*. Wiley.

83 Pawar, S.P., Bhingardive, V., Jadhav, A., and Bose, S. (2015). An efficient strategy to develop microwave shielding materials with enhanced attenuation constant. *RSC Advances* 5 (109): 89461–89471.

84 Al-Saleh, M.H. and Sundararaj, U. (2009). Electromagnetic interference shielding mechanisms of CNT/polymer composites. *Carbon* 47 (7): 1738–1746.

85 González, M., Pozuelo, J., and Baselga, J. (2018). Electromagnetic shielding materials in GHz range. *The Chemical Record* 18: 1000–1009.

86 Li, X., Yi, H., Zhang, J. et al. (2013). Fe_3O_4–graphene hybrids: nanoscale characterization and their enhanced electromagnetic wave absorption in gigahertz range. *Journal of Nanoparticle Research* 15 (3): 1472.

87 Feng, J., Pu, F., Li, Z. et al. (2016). Interfacial interactions and synergistic effect of CoNi nanocrystals and nitrogen-doped graphene in a composite microwave absorber. *Carbon* 104: 214–225.

88 Wang, G., Gao, Z., Wan, G. et al. (2014). High densities of magnetic nanoparticles supported on graphene fabricated by atomic layer deposition and their use as efficient synergistic microwave absorbers. *Nano Research* 7 (5): 704–716.

89 Quan, L., Qin, F., Estevez, D. et al. (2017). Magnetic graphene for microwave absorbing application: towards the lightest graphene-based absorber. *Carbon* 125: 630–639.

90 Zhu, Y., Pan, Y., Yang, Z. et al. (2019). Ruderman–Kittel–Kasuya–Yosida mechanism for magnetic ordering of sparse Fe adatoms on graphene. *The Journal of Physical Chemistry C* 123 (7): 4441–4445.

91 Khurana, G., Kumar, N., Kotnala, R. et al. (2013). Temperature tuned defect induced magnetism in reduced graphene oxide. *Nanoscale* 5 (8): 3346–3351.

92 Rao, C., Matte, H.R., Subrahmanyam, K., and Maitra, U. (2012). Unusual magnetic properties of graphene and related materials. *Chemical Science* 3 (1): 45–52.

93 Agarwal, M. and Mishchenko, E. (2017). Long-range exchange interaction between magnetic impurities in graphene. *Physical Review B* 95 (7): 075411.

94 Zhu, T., Shen, W., Wang, X. et al. (2019). Paramagnetic CoS_2@ MoS_2 core–shell composites coated by reduced graphene oxide as broadband and tunable high-performance microwave absorbers. *Chemical Engineering Journal* 378: 122159.

5

MXenes as EMI Shielding Materials

5.1 Introduction

Transition metal carbides/nitrides/carbonitrides (MXenes) are the fastest growing family of two-dimensional (2D) materials owing to their unique intrinsic properties, including excellent metallic conductivity caused by excess electron density near the Fermi level (E_F), hydrophilicity arising from the presence of surface functional groups (—OH, —F, —Cl, =O, etc.), and the tunability of surface chemistry [1–3]. This excellent metallic conductivity coupled with their hydrophilic nature has extended the applications of MXene into diverse fields, including electromagnetic interference (EMI) shielding [4–8], electrode materials for batteries and supercapacitors [9, 10], optoelectronics [11], sensors [12], fluorescence [13, 14], thermal heaters [15], dielectric materials [16], triboelectric nanogenerators [17], and thermoelectrics [18].

2D MXenes, which have the general formula $M_{n+1}X_nT_x$, are derived from parent three-dimensional (3D) layered MAX phases by selective chemical etching of the "A" elements from group 13 or 14, i.e. Al or Si (in the form of Al^{3+} and Si^{4+}), under aqueous acidic conditions (Figure 5.1a) [22, 23]. Discovered by Hans Nowotny's group in Austria in the 1960s [24] and explored by Barsoum and El-Raghy in the mid-1990s in the United States [25], more than 100 pure MAX phase materials have been reported so far. With the addition of solid solutions and double transition metal structures, this number can be greatly extended [26]. All the known MAX phases in the space group $P6_3/mmc$ possess a layered hexagonal structure. The M layers are nearly closely packed, whereas the X atoms occupy the octahedral sites. The A layer atoms separate the repeated $M_{n+1}X_n$ layers and develop metallic bonds (M—A) with the outer M layer atoms (Figure 5.1b). In contrast, the M—X bonds are covalent/ionic/metallic in nature and thus much stronger than the M—A bonds. Depending on the number of M layers separated by A layers, the MAX phases are classified as 211, 312, and 413 phases. Ti- and Al-based $Ti_{n+1}AlC_n$ MAX phases have a constant lattice parameter ($a = \sim 3$ Å), whereas the vertical plane stacking distance varies with the number of M layers, with $c = \sim 13$, ~ 18, and ~ 23–24 Å for the 211, 312, and 413 phases, respectively [27]. The unit cells consist of M_6X octahedra interleaved with layers of A elements [28]. As transition metal ions generally have a coordination number of 6, it is commonly assumed that the transition metals in

Two-Dimensional Materials for Electromagnetic Shielding, First Edition.
Chong Min Koo, Pradeep Sambyal, Aamir Iqbal, Faisal Shahzad, and Junpyo Hong.

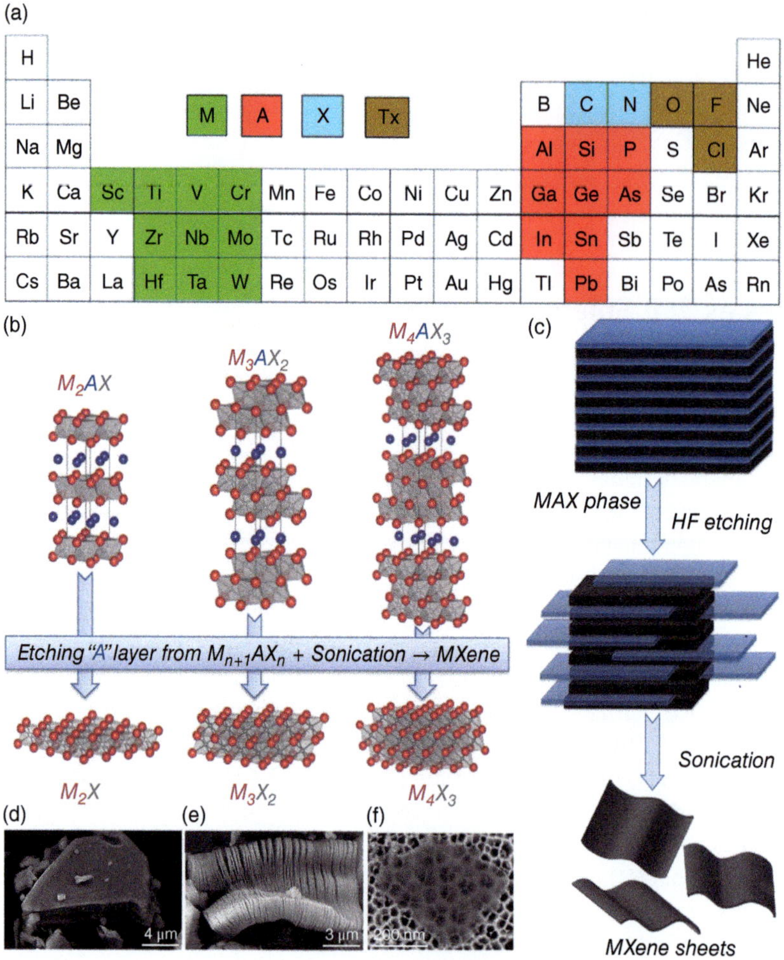

Figure 5.1 (a) Elements in MAX phase materials and corresponding MXenes. Source: Malaki et al. [19]. Reproduced with permission of the Royal Society of Chemistry. (b) Structures of various MAX phases and corresponding MXenes. Source: Naguib et al. [20]. Reproduced with permission of Wiley-VCH. (c) Schematic of the synthesis of MXenes from MAX phases. Scanning electron microscopy (SEM) images of (d) a Ti_3AlC_2 particle and (e) the accordion-like multilayer structure of $Ti_3C_2T_x$ after removal of the Al atoms from Ti_3AlC_2 by HF etching. Source: Naguib et al. [21]. Reproduced with permission of the American Chemical Society. (f) Exfoliated single-layer $Ti_3C_2T_x$ MXene sheet on an anodized aluminum oxide disk. Source: Shahzad et al. [4]. Reproduced with permission of AAAS, Science.

MXenes form six chemical bonds with neighboring X atoms and surface terminations (T_x: OH, O, and F) to obtain $M_{n+1}X_nT_x$, e.g. M_2XF_2, $M_2X(OH)_2$, and M_2XO_2 [20, 29]. Most MAX phases can be converted into corresponding 2D MXenes, thus drastically increasing the total number of MXene materials [30].

MAX phases can be synthesized with varying compositions of M, A, and X elements in the form of a perfect structure or a solid solution, which adds numerous new candidates for MAX phases [20]. Depending on the M and X elements, MAX phases can be categorized as single-transition-metal MXenes, double-transition-metal (M′ and M″) MXenes (with ≥3 M layers), and double-X solid-solution MXenes with varying C and N ratios (carbonitrides). The two different transition metals in double-transition-metal MXenes can adopt two arrangements depending on their planar configurations: (i) a solid solution, in which the M sites are occupied by M′ and M″ atoms in a random manner and (ii) a perfectly ordered distribution, with M′ and M″ atoms in out-of-plane (o-MAX) and in-plane (i-MAX) configurations [31, 32]. Adding a fourth element to the MAX phase yields a quaternary phase or a solid solution, e.g. Ti_3AlCN and $Ti_3(Si,Al)C_2$, which can also be transformed into MXenes, adding more members to the MXene family. So far, $Mo_2TiC_2T_x$, $Mo_2Ti_2C_3T_x$, and $Cr_2TiC_2T_x$ double-transition-metal MXenes derived from the corresponding o-MAX phases have been experimentally synthesized. Interestingly, these MXenes with two different metals exhibit electronic properties far different from those of the single-transition-metal MXene $Ti_3C_2T_x$. A different electronic conduction mechanism results in these materials being semiconductor-like MXenes, in contrast to conventional $Ti_3C_2T_x$ and other MXenes, which have conductivities similar to those of metals.

The removal of the A layers from ternary carbides/nitrides yields the corresponding MXenes (Figure 5.1c). In contrast to other 2D materials, such as graphene and black phosphorus (BP), it is difficult to separate the A layer from the MAX phase by traditional exfoliation methods owing to the strong M—A metallic bonds [22, 33, 34]. Fortunately, the strong M—X bonds (ionic/covalent mixed) favor breaking the M—A bonds chemically. To selectively etch the A layers of atoms, processes using several different mixtures of acids and salts have been reported, including hydrofluoric acid (HF) etching, in situ formed HF etching, molten fluoride salt etching, and fluorine-free etching [35]. In 2011, Naguib et al. opened up the area of MXenes by achieving the selective chemical etching of the Al atoms from the Ti_3AlC_2 MAX phase (Figure 5.1d) in an aqueous solution of HF for two hours, followed by exfoliating the accordion-like multilayer-stacked $Ti_3C_2T_x$ MXene (Figure 5.1e) into single- or few-layer MXene flakes through intercalation of cations or mild sonication (Figure 5.1f) [22]. The chemical etching process to convert MAX into MXenes can be described as follows:

$$Ti_3AlC_2 + 3HF = Ti_3C_2 + AlF_3 + 3/2H_2 \tag{5.1}$$

$$Ti_3C_2 + 2H_2O = Ti_3C_2(OH)_2 + H_2 \tag{5.2}$$

$$Ti_3C_2 + 2HF = Ti_3C_2F_2 + H_2 \tag{5.3}$$

As shown in Eqs. (5.1)–(5.3), the Al atoms are stripped via a reaction with a strong acid (HF) to form Ti_3C_2. Then, the bare, exposed metal (Ti) layers react with H_2O and HF in the etching solvent to form corresponding compounds with –OH and –F terminations. The unstable –OH terminations are converted to –O terminal groups ($-OH + -OH = -O + H_2O$) after gaining an electron [36, 37].

Subsequently, this method was extended to the synthesis of many other MXenes, e.g. Ti_2C, V_2C, Nb_2C, Ti_3CN, and Ta_4C_3, from their corresponding MAX phases [21, 38]. The optimal etching conditions varied slightly depending on the MAX precursor and the number of M layers. M layers with more valence electrons or higher atomic numbers as well as higher numbers of M layers result in stronger M—A bonds. Therefore, stronger etching conditions (etching time, temperature, and HF concentration) are required to synthesize MXenes with larger n values in $M_{n+1}X_nT_x$ [21, 31, 38].

The HF etching method does not yield mono- or few-layer MXenes, as the obtained multilayered structure has strong interlayer bonding. Following the same protocol, Mashtalir et al. delaminated the multilayered structure to few-layer MXenes by introducing intercalated ions and/or molecules [39, 40]. In an extra step after HF etching, the multilayer MXenes were intercalated with dimethyl sulfoxide (DMSO) via ultrasonication in water. The induced swelling subsequently weakened the interlayer bonding between the MXene layers. Naguib et al. intercalated multilayer $Ti_3C_2T_x$ MXene with tetrabutylammonium hydroxide (TBAOH) to obtain stable dispersions of mono- to few-layer MXenes [41]. Other studies employed urea [42], tetramethylammonium hydroxide (TMAOH) [35, 43], and tetrapropylammonium hydroxide (TPAOH) [44] as intercalants in multilayered $Ti_3C_2T_x$ MXene to obtain single- or few-layer MXene sheets.

The HF-assisted etching method is a two-step process involving exfoliation and intercalation. Although this method is fast and simple, it provides low reaction yields and the produced MXene sheets are quite defective with overetched Ti layers. Moreover, the direct use of HF is hazardous to the users' health. Therefore, Ghidiu et al. reported an alternative method without HF (Eq. (5.4)) [45]. The use of 6 M HCl with LiF salt simultaneously improved the synthesis procedure and the yield of MXene. The obtained clay-like $Ti_3C_2T_x$ MXene showed outstanding performance in energy storage applications. The chemical process involved can be summarized as follows:

$$2Ti_3AlC_2 + 6LiF + 6HCl = 2Ti_3C_2 + Li_3AlF_6 + AlF_3 + 3LiCl + 3H_2 \tag{5.4}$$

Shahzad et al. slightly modified the etching recipe to develop the minimal intensive layer delamination (MILD) method. The Ti_3AlC_2/LiF/HCl molar ratio was changed from 1.0 : 5.0 : 11.7 to 1.0 : 7.5 : 23.4, which made the delamination process much easier without the need for intercalation and sonication [4]. The obtained high-quality MXene nanosheets showed an outstanding electrical conductivity and EMI shielding effectiveness (SE) of 4665 S cm^{-1} and 92 dB, respectively, at a thickness of 45 μm. The MXenes synthesized by the LiF/HCl method do not require an extra intercalation step because the Li^+ ions act as an intercalant, assisting the delamination of MXene sheets with a larger interlayer distance (d-spacing). In addition, the obtained MXene sheets are less defective and hence more conductive and oxidation-resistant. Therefore, this route has been adopted in many recent studies, even for carbonitride Ti_3CNT_x MXene [8, 46]. Table 5.1 summarizes the different etching conditions for a range of reported MXenes.

In addition to these methods, some mild etchants have been reported for the synthesis of $Ti_3C_2T_x$ MXene. Halim et al. used NH_4HF_2 to etch Ti_3AlC_2, where insertion of the NH^{4+} ions into the MXene layers resulted in a larger d-spacing and an enhanced degree of delamination [59]. Similarly, monocation fluoride (LiF, NaF, KF, and NH_4F) [60] and bifluoride salts ($NaHF_2$, KHF_2, and NH_4HF_2) were also used for

Table 5.1 Conditions for the synthesis of various MXenes.

MAX precursor	MXene	Etching conditions	Temperature (°C)	Time (hours)	Delamination conditions	Conductivity ($S\,cm^{-1}$)	References
Ti_3AlC_2	$Ti_3C_2T_x$	50% HF	RT	2	Sonication	~3 000–15 000 depending on synthesis, post-etching, and films' fabrication conditions	[22]
		50% HF	RT	18	DMSO		[47]
		5% HF	RT	24	DMSO + TMAOH		[35]
		10% HF		18			
		30% HF		5			
		49% HF + 9 M HCl	N/A	24	LiCl		[48]
		6 M HCl + 5 M LiF	40	45	N/A		[45]
		9 M HCl + 1.9 M LiF	35	24	Sonication		[49]
		9 M HCl + 3.8 M LiF	35	24	Sonication		[50]
		9 M HCl + 7.5 M LiF	35	24	Not required		[4]
		12 M HCl + 3 M LiF	35	48	Not required		[51]
		2.25 M NH_4F	150	24	N/A		[52]
		HF + HCl	RT	24	LiCl	8 570	[53]
Ti_3AlCN	Ti_3CNT_x	LiF + HCl	RT	24	N/A	1 150–2 712	[8]
Ti_2AlC	Ti_2CT_x	10% HF	RT	10	N/A	1 600–2 000	[54]
V_2AlC	V_2CT_x	28 M HF	RT	45	TBAOH	998–1 560	[55]
Nb_2AlC	Nb_2CT_x	50% HF	55	48	Isopropylamine	5	[56]
Nb_4AlC_3	$Nb_4C_3T_x$	50% HF	50	7 d	TMAOH	75	[53]
$Mo_2Ti_2AlC_3$	$Mo_2Ti_2C_3T_x$	10% HF + 10% HCl	40–50	40–96	N/A	227–297.9	[4]

(Continued)

Table 5.1 (Continued)

MAX precursor	MXene	Etching conditions	Temperature (°C)	Time (hours)	Delamination conditions	Conductivity (S cm^{-1})	References
Mo_2TiAlC_2	$Mo_2TiC_2T_x$	10% HF + 10% HCl	40–50	40–48	N/A	49–119.7	
Mo_2Ga_2C	Mo_2CT_x	14 M HF	55	160	TBAOH	0.1–1	[57]
$(Mo_{2/3}Sc_{1/3})_2AlC$	$Mo_{1.33}CT_x$	48% HF	RT	10	TBAOH	296.74	[58]
$Ti_{0.4}Nb_{1.6}AlC$	$Ti_{0.4}Nb_{1.6}CT_x$	LiF + HCl	35	48	N/A	55	[53]
$Ti_{0.8}Nb_{1.2}AlC$	$Ti_{0.8}Nb_{1.2}CT_x$					518	
$Ti_{1.2}Nb_{0.8}AlC$	$Ti_{1.2}Nb_{0.8}CT_x$					867	
$Ti_{1.6}Nb_{0.4}AlC$	$Ti_{1.6}Nb_{0.4}CT_x$					1 370	
$Nb_{1.6}V_{0.4}AlC$	$Nb_{1.6}V_{0.4}CT_x$	HF	35	48	TMAOH	19	
$Nb_{1.2}V_{0.8}AlC$	$Nb_{1.2}V_{0.8}CT_x$					51	
$Nb_{0.8}V_{1.2}AlC$	$Nb_{0.8}V_{1.2}CT_x$					108	
$Nb_{0.4}V_{1.6}AlC$	$Nb_{0.4}V_{1.6}CT_x$					349	

the synthesis of $Ti_3C_2T_x$ MXene [61]. The chemical procedure using bifluoride salts can be generalized as follows:

$$Ti_3AlC_2 + XHF_2 = Ti_3C_2 + AlF_3 + X_aAlF_b + H_2 \quad (5.5)$$

Yang et al. reported the electrochemical etching of Ti_3AlC_2 MAX for the synthesis of fluorine-free $Ti_3C_2T_x$ MXene with enhanced corrosion resistance [62]. However, this process yielded multilayered MXenes, and the reaction could not be feasibly scaled. Similarly, a few studies have revealed the hydrothermal synthesis of $Ti_3C_2T_x$ and Nb_2CT_x MXenes with a less-toxic $NaBF_4$ + HCl etchant [63]. The synthesized MXenes possessed larger Brunauer–Emmett–Teller (BET) surface areas owing to slow release during etching, which was unfortunately accompanied by a higher and faster degree of oxidation. However, such MXenes may be beneficial depending on the intended application. For example, the oxidized MXenes showed better performance for the adsorption of methylene blue and methyl orange dyes. Similarly, highly fluorescent MXene quantum dots synthesized by a hydrothermal method showed huge potential in bioimaging and sensing applications [13, 14].

As the synthesis process strongly affects the physiochemical, electrical, and mechanical properties of MXenes, the synthesis route should be chosen depending on the intended area of application. For example, MXenes synthesized by the LiF/HCl method were more conductive than those obtained by HF etching and hence showed better EMI SE [64]. The properties of MXenes are also affected by several extrinsic factors such as the MAX precursor [65], MXene synthesis conditions [46], post-etching treatments [35], ultrasonication [19], surface terminations [66], storage environment and temperature [67, 68], and defect density [69]. These variables can be used to tune the conductivity of MXene films from

$1\,S\,cm^{-1}$ to as high as ~15 000 $S\,cm^{-1}$ [70, 71]. The MAX precursor was found to influence the electrical conductivity of the resulting $Ti_3C_2T_x$ MXene, with the graphite-based carbon precursor Ti_3AlC_2 resulting in a highly conductive $Ti_3C_2T_x$ MXene (~4400 $S\,cm^{-1}$) with enhanced stability (time constant of 10.1 days) [65]. Therefore, MXenes are emerging as ideal 2D materials for applications that demand high electrical conductivity. Additionally, the tunable surface chemistry of MXenes ensures the formation of stable dispersions in various solvents for different liquid-manufacturing techniques, e.g. spray coating, spin casting, dip coating, inkjet printing, and interfacial assembly [72, 73]. Recently, Kim et al. [72] reported nonpolar organic dispersions of functionalized $Ti_3C_2T_x$ MXenes, thus expanding the development of MXene hybrids and composites in solution [22, 35]. Therefore, for MXenes, the flake size [46], electrical conductivity [4, 71], surface terminations [36, 74–76], and number and type of defects [69] can be customized by controlling the aforementioned parameters.

5.2 MXenes for EMI Shielding

Since the discovery of the first MXene ($Ti_3C_2T_x$) in 2011 [22], more than 30 different MXenes have been experimentally synthesized, with an even larger number predicted computationally for future discovery [30]. MXenes can play a constructive role in EMI shielding, as they possess all the perquisite features of an efficient shielding material, i.e. a 2D nature with a large specific surface area, low density, flexibility, and, most importantly, easy processability, along with outstanding metallic electrical conductivity [4]. For example, a $Ti_3C_2T_x$ MXene with an electrical conductivity of ~4665 $S\,cm^{-1}$ delivered an outstanding EMI SE of 92 dB at a thickness of 45 μm, which was higher than those of other synthetic materials with comparable thicknesses [4, 77]. Polymeric composites with biocompatible sodium alginate (SA) resulted in an absolute shielding effectiveness (specific shielding effectiveness [SSE]/t; EMI SE normalized by the density and thickness of shield) of more than 30 000 $dB\,cm^2\,g^{-1}$, revealing a promising potential for lightweight EMI shielding applications. Interestingly, the effect of conductivity on the EMI shielding performance was well demonstrated in a recent study, in which 16 different types of MXenes were compared and those ($Ti_3C_2T_x$ in this case) with the highest conductivity exhibited the best EMI shielding capability [53]. Very recently, the scalable manufacturing of freestanding MXene films with dimensions of 1 m (length) by 10 cm (width) was demonstrated. The film with a thickness of 940 nm exhibited an outstanding tensile strength of ~570 MPa, an electrical conductivity of ~15 100 $S\,cm^{-1}$, and an EMI shielding performance of ~50 dB, which surpasses the performance of all existing 2D materials produced to date [71]. The unique properties of MXenes make them ideal for EMI shielding, as high electrical conductivity is pivotal for efficient EMI shielding materials.

The fast growth of portable, smart, and flexible electronics demands EMI shielding materials with ultralow thicknesses, lightweights, and exceptional shielding performance. In this chapter, we highlight the development of 2D MXenes in the field of advanced EMI shielding materials. In particular, the synthesis of different MXenes, their electrical and chemical properties, and the structure–property relationships are described in detail.

Unlike graphene, MXenes are a large family comprising various transition metals that form structures with several atom layer thicknesses. Monolayer MXenes have been reported to be as transparent as single-layer graphene [78]. During the etching of the A layers, hydroxyl and fluorine groups are attached to the A sites, forming metal–oxygen and metal–halide bonds that enhance the absorption of electromagnetic waves (EMWs) in layered MXenes by dipolar polarization losses. Moreover, the tunable surface chemistry of MXenes favors the development of composites with various designed structures, including thin compact laminates, layer-by-layer (LbL) assemblies with reinforcing constituents, porous foams, and segregated nanostructures, as discussed in a recent study [79]. Since the first report on the EMI shielding properties of $Ti_3C_2T_x$ MXene by Shahzad et al. [4], a number of interesting studies have further enhanced the intrinsic properties of $Ti_3C_2T_x$ and explored other MXenes [5–8].

With the growing demand for summarized knowledge, it is important to highlight the research interest in MXene materials for EMI shielding applications, which has become a hot topic in the area of 2D materials. In this chapter, MXenes are classified based on their structural design to better understand the shielding parameters, their effective roles, and the involved mechanisms. For this purpose, the structural design morphologies were classified as compact and laminated MXene structures, MXene hybrids and composites in LbL assemblies, porous foams and aerogels, and segregated structures. In addition, comparisons are made to show that the performance of MXenes surpasses that of conductive metals and carbon-based conductive and nonmagnetic materials. This chapter provides a comprehensive summary of the EMI shielding properties of highly conducting 2D MXenes, and highlights the future challenges with their prospective solutions in the domains of MXenes.

5.2.1 MXene Laminate Films

In 2016, as a new application area for MXenes, Shahzad et al. revealed the outstanding EMI shielding of $Ti_3C_2T_x$, along with $Mo_2TiC_2T_x$ and $Mo_2Ti_2C_3T_x$, laminate films [4]. Each MXene was synthesized from its corresponding Al-based MAX phase and collected as an aqueous dispersion consisting of delaminated 2D MXene flakes (Figure 5.2a). Vacuum-assisted filtration of the aqueous dispersion resulted in micrometer-thick and uniformly aligned MXene laminate films (Figure 5.2b). As an advantage, the hydrophilic surface of $Ti_3C_2T_x$ allowed the fabrication of polymer composites with biocompatible SA to improve the mechanical properties and environmental stability of the pure MXenes, as shown by the cross-sectional scanning electron microscopy (SEM) and high-resolution transmission electron microscopy (TEM) images in Figure 5.2c,d.

$Ti_3C_2T_x$ showed a much higher electrical conductivity (~4665 S cm^{-1}) than $Mo_2TiC_2T_x$ and $Mo_2Ti_2C_3T_x$ (~120 and ~300 S cm^{-1}, respectively; Figure 5.2e). At a thickness of only 2.5 μm, the total EMI SE values (SE_T; sum of the shielding due

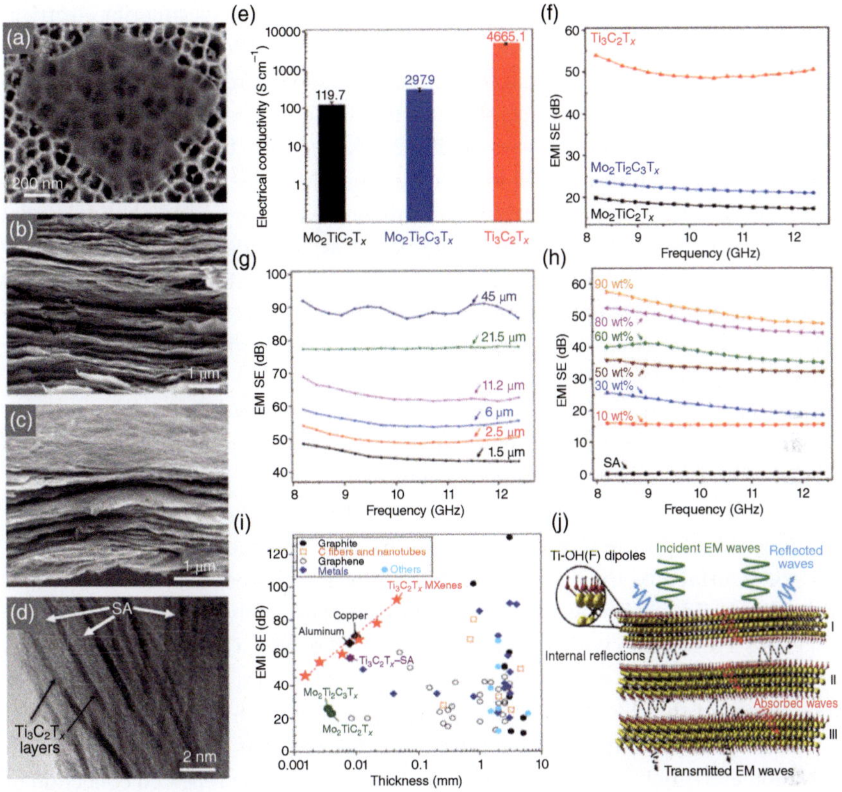

Figure 5.2 SEM images of (a) exfoliated MXene flakes and cross-sectional views of (b) a bare MXene film and (c) an MXene–SA composite film. (d) TEM image of an MXene–SA composite film. (e, f) Electrical conductivity and EMI SE of $Mo_2TiC_2T_x$, $Mo_2Ti_2C_3T_x$, and $Ti_3C_2T_x$. (g) EMI SE of $Ti_3C_2T_x$ MXene films at different thicknesses. (h) EMI SE of $Ti_3C_2T_x$–SA composite films at a thickness of 9 µm. (i) Comparison of EMI SE for MXene films and their composites with other reported materials. (j) Proposed EMI shielding mechanism in compact $Ti_3C_2T_x$ MXene laminate films. Source: Shahzad et al. [4]. Reproduced with permission of American Association for the Advancement of Science.

to absorption [SE_A] and reflection [SE_R]) of $Ti_3C_2T_x$, $Mo_2TiC_2T_x$, and $Mo_2Ti_2C_3T_x$ were 54, 20, and 24 dB, respectively, in the X-band frequency range (8.2–12.4 GHz; Figure 5.2f). The EMI SE of $Ti_3C_2T_x$ increased with the thickness and reached 92 dB at a thickness of 45 µm (Figure 5.2g). The enhanced EMI SE of $Ti_3C_2T_x$ was ascribed to its higher electrical conductivity, as the experimental results were consistent with the calculations based on Simon's equation. The excellent electrical conductivity of $Ti_3C_2T_x$ MXene favored the realization of highly conductive $Ti_3C_2T_x$-SA composites, and a 9 µm thick composite film with 90 wt% $Ti_3C_2T_x$ MXene showed an EMI SE of 57 dB (Figure 5.2h) with an SSE/t value of 30 830 dB cm^2 g^{-1}, which is the highest value reported thus far. Thus, the $Ti_3C_2T_x$ MXene outperformed all other synthetic materials, even highly conductive metals such as Ag and Cu, in terms of EMI SE (Figure 5.2i).

The excellent EMI SE of $Ti_3C_2T_x$ MXene was attributed to its superior electrical conductivity and laminate structure (Figure 5.2j). For EMWs incident on the surface of conductive MXene films, a portion of the EMWs is reflected because of the impedance mismatch between the MXene with abundant charge carriers (free electrons) and air. The remaining waves enter the MXene layered structure, where each layer individually acts as a shield to attenuate the EMW energy through eddy current and ohmic losses. Surface terminations (metal–oxygen and metal–halide bonds) cause polarization losses owing to dipole polarization under an alternating electromagnetic field, thus increasing the absorption of the EMWs inside the layered architecture. A subsequent theoretical study used the Fresnel formula to validate the EMI shielding performance of conductive $Ti_3C_2T_x$ MXene, thus prompting further exploration of 2D MXenes [80].

Yun et al. fabricated electrically thin (thickness smaller than the skin depth) large-area $Ti_3C_2T_x$ MXene films by the interfacial assembly of monolayer MXene flakes to study the effect of multiple reflections (Figure 5.3a) [78]. The total EMI SE is the sum of the shielding contributions from reflection, absorption, and multiple reflections (SE_M). In the multiple-reflection mechanism, the waves reflected from the back interface of the material undergo repeated reflections and absorption within the shield until transmitted or attenuated completely. Overall, this mechanism yields a lower SE_T with a lower SE_R and a higher SE_A. Generally, it is believed that the role of multiple reflections is negligible when the total EMI SE is ≥15 dB or the thickness of the shield is larger than the skin depth, which is the thickness of the shielding material at which the intensity of the incident EMWs is decreased to $1/e$. Nanometer-thick multilayer films were fabricated by the repeated collection of assembled monolayer films, and a larger thickness was obtained by stacking these films (Figure 5.3b). Interestingly, this technique could also be used to collect the assembled MXene films with controlled thicknesses on any substrate of choice. An assembled monolayer $Ti_3C_2T_x$ MXene film with a thickness of 2.3 nm had a transparency of 90%, similar to that of monolayer graphene, which is advantageous for flexible and transparent electronics. The uniformity of the mono- and multilayer films was confirmed by the continuous increase in absorbance and decrease in sheet resistance as a function of thickness (Figure 5.3c).

As EMI shielding is proportional to the electrical conductivity and thickness, SE_T gradually increased with the number of stacked layers, as shown in Figure 5.3d. An assembled $Ti_3C_2T_x$ MXene film with 24 layers and a thickness of 55 nm had an EMI SE of 20 dB, which is the minimal requirement for commercial applications. Moreover, owing to the nanometer thickness and low density, the highest ever SSE/t value of 3.89×10^6 dB cm^2 g^{-1} was recorded for these annealed assembled films, indicating that MXenes are applicable as ultra-lightweight EMI shielding materials.

To investigate the role of multiple reflections in films with small thicknesses, the EMI SE values of electrically thin and electrically thick (thickness larger than the skin depth of 7.86 μm) were determined as a function of thickness and compared with the values determined using two theoretical models: Simon's formula [81], which does not consider multiple reflections, and the transfer matrix method [82–84] that considers multiple reflections. The experimental SE_T, SE_R, and SE_A

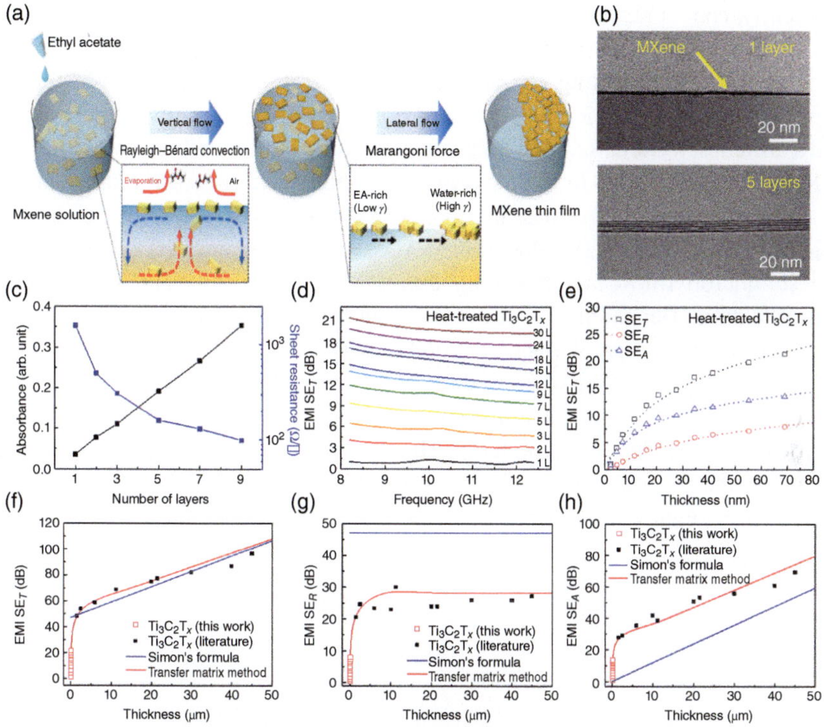

Figure 5.3 (a) Schematic of the fabrication process of interfacial self-assembled MXene thin films. (b) Cross-sectional TEM images of monolayer (1L) and five-layer (5L) films. (c) Sheet resistance and optical absorbance at 550 nm of assembled MXene films with various numbers of stacked layers. (d) EMI SE_T in the X-band frequency range of assembled MXene films with various numbers of layers annealed at 400 °C. (e) Comparison of experimental SE_T, SE_R, and SE_A values of electrically thin MXene films with those calculated using the transfer matrix method. (f–h) Comparison of experimental SE_T, SE_R, and SE_A values (red open dots) with the simulated values (red line: transfer matrix method, blue line: Simon's formula) and literature values [4] (black dots). Source: (c–f) Yun et al. [78]. Reproduced with permission of Wiley-VCH.

values corresponded well with the values calculated using the transfer matrix method (Figure 5.3e). Moreover, both the electrically thin and thick MXene films showed absorption-dominant shielding mechanisms ($SE_A > SE_R$). As shown in Figure 5.3f, the SE_T values for the electrically thick samples were consistent with Simon's assumptions (blue line), whereas this formula was not effective for the electrically thin films, revealing that multiple reflections occur at lower thicknesses. In this case, Simon's formula overestimates the SE_R values (Figure 5.3g) and underestimates the SE_A values (Figure 5.3h) by neglecting the multiple reflections. In contrast, the values calculated using the transfer matrix method (red line) were in good agreement with the experimental values of SE_T, SE_R, and SE_A of nanometer-thin MXene films. This was the first experimental study to differentiate between electrically thin and electrically thick EMI shielding materials.

Recently, Han et al. also validated the role of multiple reflections in nanometer-thin films fabricated by spray coating and spin casting. In addition, the authors reported the EMI shielding characteristics of 16 different MXene laminates by fabricating submicron-thick films using vacuum filtration [53]. In this systematic study, MXenes were classified based on the number of atomic layers, i.e. M_2X, M_3X_2, and M_4X_3, single- or double-transition metals, and solid solutions to study the effect of elemental composition, the arrangement of transition metal layers, and the layer structure. All the studied MXenes showed an EMI SE of >20 dB at a submicron thickness, indicating that MXenes have a strong potential for EMI shielding. Depending on the composition and the number of M layers, there was a large range of electrical conductivity values from 5 to ~8500 S cm^{-1} (Figure 5.4a). $Ti_3C_2T_x$ MXene showed the highest shielding efficiency, with an EMI SE of 21 dB recorded for a 40 nm thick spray-coated film, owing to the superior electrical conductivity of this material. This finding also emphasized the practical role of electrical conductivity in defining the EMI shielding ability of a material (Figure 5.4b). Figure 5.4c summarizes the EMI SE values of all the MXenes. It

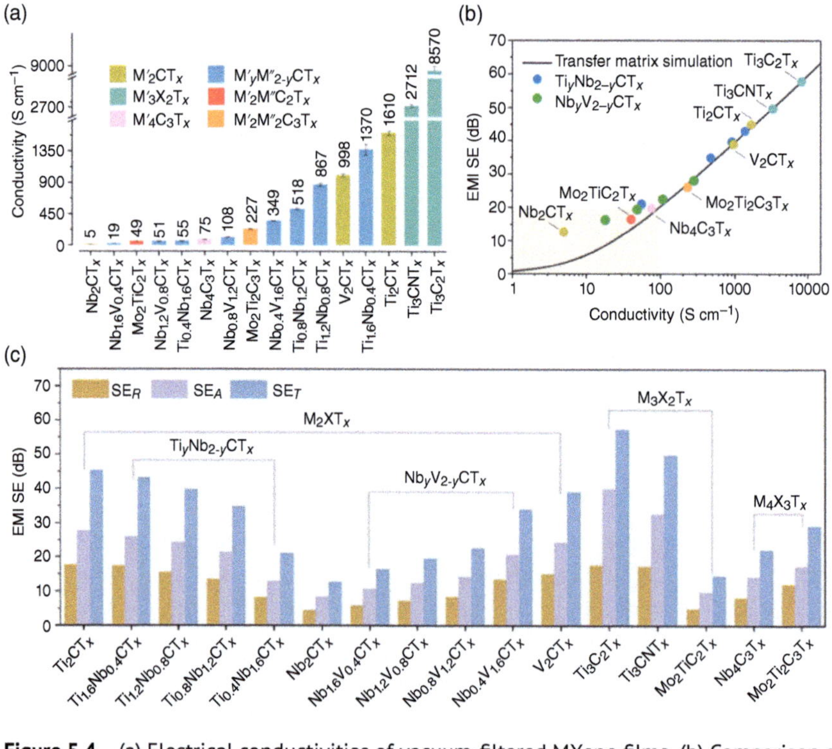

Figure 5.4 (a) Electrical conductivities of vacuum-filtered MXene films. (b) Comparison of EMI SE values of vacuum-filtered MXene films with a thickness of ~5 μm at 10 GHz with the transfer matrix simulation, showing the relationship between EMI SE and the electrical conductivity of MXenes. (c) Average EMI SE values (SE_R, SE_A, and SE_T) of various MXene films (5 ± 0.3 μm thick) in the 8.2–12.4 GHz range, showing the reflection and absorption contributions. Source: Han et al. [53]. Reproduced with permission of the American Chemical Society.

was found that the EMI shielding performance of the MXenes could be tailored by controlling the transition metal composition, the number of layers, and the mixed double-transition-metal solid solution, which emphasizes that MXenes are promising materials for a wide range of EMI shielding applications.

The synthesis conditions strongly affect the properties of MXenes. Zhang et al. reported the highest electrical conductivity for a $Ti_3C_2T_x$ MXene synthesized using a modified LiF + HCl method [71]. In this study, the Ti_3AlC_2 MAX powder was sieved to collect large multilayer flakes with an average size of >10 μm (Figure 5.5a). The corresponding $Ti_3C_2T_x$ MXene with a large average flake size of 10 ± 2.1 μm formed lyotropic liquid crystals. With high viscosity, the enhanced flowability under shear stress aided in the fabrication of MXene films by blade coating with sufficient shear stress on a Celgard membrane (Figure 5.5b). The MXene film with a larger flake size and a thickness of 940 nm exhibited excellent mechanical properties (Figure 5.5c–f) owing to the high compactness of the large flakes. The tensile strength of this film (568 ± 24 MPa) was over 30 times higher than that of pure MXene films (~20 MPa), and Young's modulus reached 20.6 ± 3.1 GPa. Owing to the high alignment of the larger flakes and reduced interflake resistance, the MXene films showed an extraordinary electrical conductivity of 15 100 $S\,cm^{-1}$ after annealing at 200 °C for six hours, which is the highest reported value for an experimentally synthesized MXene (Figure 5.5g). The highly conductive MXene film with a thickness of 940 nm showed a promising EMI SE of ~48 dB in the X-band frequency range (Figure 5.5h). This synthesis route ensures the scalable production of $Ti_3C_2T_x$ MXene with outstanding EMI shielding properties for real applications.

Similarly, He et al. reported the electrical properties of $Ti_3C_2T_x$ MXenes synthesized by two different routes, namely, HF (40%) etching at room temperature for 24 hours and LiF + HCl treatment at 40 °C for 16 hours [64]. The HF etching route gave an accordion-like multilayer MXene (M-$Ti_3C_2T_x$), whereas the LiF + HCl method produced an ultrathin MXene (U-$Ti_3C_2T_x$). The HF etching method weakened the M—A bonds but not enough to delaminate the multilayered structure into separate layers. In the LiF + HCl method, in situ synthesized HF etched the Al layers, whereas the Li ions acted as intercalants to obtain mono- to few-layer MXenes. M-$Ti_3C_2T_x$ had more —F terminations, whereas =O terminations were predominant in U-$Ti_3C_2T_x$. Owing to its larger surface area, U-$Ti_3C_2T_x$ showed a higher electrical conductivity than M-$Ti_3C_2T_x$. This electrical conductivity difference also resulted in their composites in SiO_2 matrices, which are transparent to EMWs, having different EMI shielding properties. Nanocomposite films with a thickness of 1 mm were fabricated using each of the MXene at the same mass loading, followed by cold pressing under 5 MPa. At a similar loading of 60 wt% MXene, the U-$Ti_3C_2T_x$ composite was more conductive (4.2×10^{-3} $S\,cm^{-1}$) than the M-$Ti_3C_2T_x$ composite (6.3×10^{-5} $S\,cm^{-1}$). As a result, the nanocomposite film of U-$Ti_3C_2T_x$ at a 80 wt% loading exhibited an EMI SE of 58 dB in the X-band frequency range that was much higher than that of the M-$Ti_3C_2T_x$ composite (35 dB). These results also confirmed that conductivity was the primary factor for EMI shielding and that the same MXene can be synthesized with a range of conductivities. Moreover, the larger exposed surface area of U-$Ti_3C_2T_x$ with abundant surface terminations and point

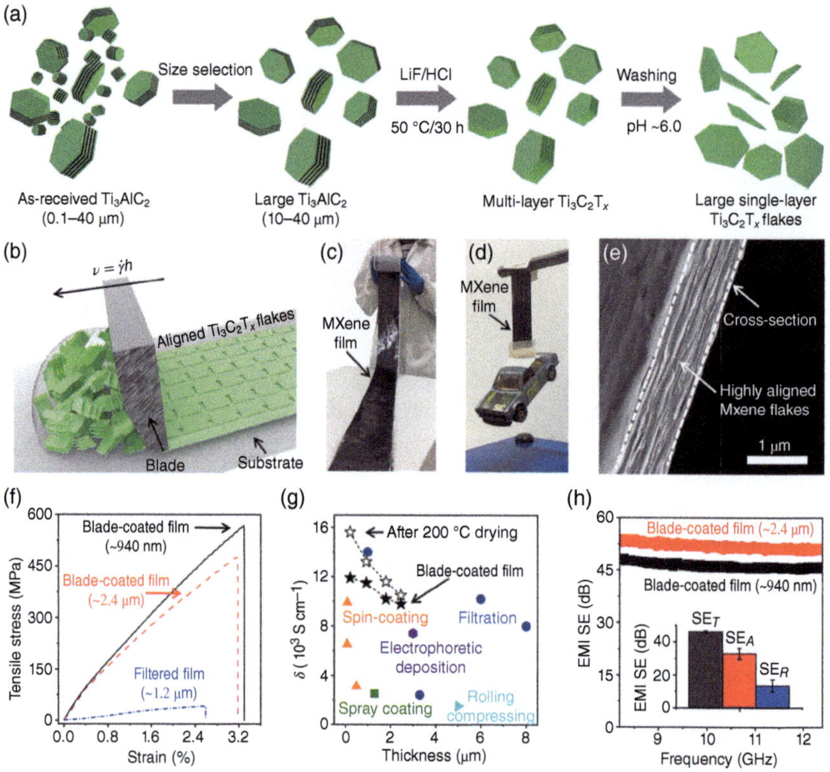

Figure 5.5 (a) Synthesis process of large $Ti_3C_2T_x$ MXene flakes from large Ti_3AlC_2 MAX particles. (b) Schematic illustration of the blade coating process. (c) Digital photograph of a 1 m × 10 cm $Ti_3C_2T_x$ film produced by blade coating large MXene flakes onto a Celgard membrane (film thickness of 940 nm). (d) Digital photograph of a piece of the blade-coated $Ti_3C_2T_x$ film (2.0 cm × 6.5 cm, thickness of 940 nm) made from large flakes lifting a ~40 g object. (e) Cross-sectional SEM image of the blade-coated film containing highly aligned large MXene flakes. (f) Typical stress–strain curves of blade-coated and vacuum-filtered films prepared from large MXene flakes. (g) Comparison of the electrical conductivities of blade-coated $Ti_3C_2T_x$ films at various thicknesses before and after drying at 200 °C for six hours with those of other reported MXenes. (h) EMI SE performance of the blade-coated ~940 nm thick $Ti_3C_2T_x$ film made from large flakes in the X-band. The inset shows the average SE_T, SE_R, and SE_A values of the film. Source: Zhang et al. [71]. Reproduced with permission of Wiley-VCH.

defects resulted in a stronger dipolar polarization mechanism and hence enhanced absorption of EMWs in the nanocomposites.

In a similar way, Li et al. studied the EMI shielding of the Ti_2AlC MAX phase and the corresponding Ti_2CT_x MXene (synthesized by the modified LiF + HCl method) using nanocomposites in a paraffin wax matrix [85]. Despite Ti_2AlC MAX ($1.37\ S\ cm^{-1}$) having a higher electrical conductivity than the corresponding MXene ($0.30\ S\ cm^{-1}$), the composites of exfoliated Ti_2CT_x MXene showed higher electrical conductivities at all mass loadings (40–100 wt%). At 40 wt%, the electrical conductivities of the Ti_2AlC/paraffin and Ti_2CT_x/paraffin composites were 5.57×10^{-9} and 1.63×10^{-8} S cm^{-1}, respectively. This difference can be attributed to the fact that the exfoliated Ti_2CT_x MXene sheets with larger surface areas aided in the efficient construction of a strong conductive network at a lower loading. Owing to the superior electrical conductivity, the delaminated Ti_2CT_x MXene exhibited a better EMI SE (70 dB) than the Ti_2AlC MAX phase (46.2 dB) at thicknesses of >0.8 mm.

The intrinsic EMI shielding properties of MXenes synthesized by different approaches vary dramatically. Very recently, Iqbal et al. reported the outstanding EMI shielding performance of a Ti_3CNT_x MXene after a mild heat treatment process [8]. Ti_3CNT_x was obtained from the Ti_3AlCN MAX phase using the typical LiF + HCl etching method. Unlike $Ti_3C_2T_x$, which was used as a reference material, the Ti_3CNT_x MXene was synthesized at room temperature. Bulk compact films of Ti_3CNT_x and $Ti_3C_2T_x$ with thicknesses of 10, 20, 30, and 40 μm were fabricated by vacuum-assisted filtration, followed by thermal annealing at 150, 250, or 350 °C for six hours under an argon atmosphere. The thermal annealing process induced greater porosity in the heat-treated Ti_3CNT_x films (Figure 5.6a) than in the $Ti_3C_2T_x$ films (Figure 5.6b). The porosity increased with increases in both the annealing temperature and the initial thickness. The compact structure of the $Ti_3C_2T_x$ films and the decreased interlayer distance (*d*-spacing) slightly improved the electrical conductivity in the measured temperature range. In contrast, the room temperature electrical conductivity of the Ti_3CNT_x film (1125 S cm^{-1}) dramatically increased to 2475 S cm^{-1} at 250 °C, whereas further increasing the temperature decreased the conductivity owing to the initiation of surface oxidation (Figure 5.6c). Interestingly, the EMI SE of the Ti_3CNT_x films was significantly improved after annealing (Figure 5.6d,e). The annealing time was also found to effectively enhance the EMI SE of the Ti_3CNT_x films (Figure 5.6f), with a 40 μm thick Ti_3CNT_x MXene film annealed at 350 °C showing an extraordinary EMI SE of 116 dB, which is much higher than that of $Ti_3C_2T_x$ (93 dB) under the same thickness and annealing conditions. This outstanding shielding performance is the best among all synthetic materials reported to date, including carbon-based materials and highly conductive Cu and Al metal foils (Figure 5.6g). The outstanding SE was governed by the induced porosity, increased electrical conductivity, moderate surface oxidation, and meta-material-like porous architecture, which favored the absorption of incident EMWs within the porous structure of Ti_3CNT_x MXene. The surface terminations also enhanced the absorption contribution by inducing dipole polarization losses. Such an improved absorption shielding efficiency enhanced the total shielding performance of the Ti_3CNT_x MXene, mitigating the effects of re-reflected EMWs that can cause secondary EMI pollution. Such a behavior is a fundamental requirement for materials to be used in highly compacted fifth-generation (5G) electronics, where a reflection-dominant shielding mechanism is not suitable. Moreover, the annealed Ti_3CNT_x films retained sufficient mechanical properties for use in portable, lightweight, and flexible electronics.

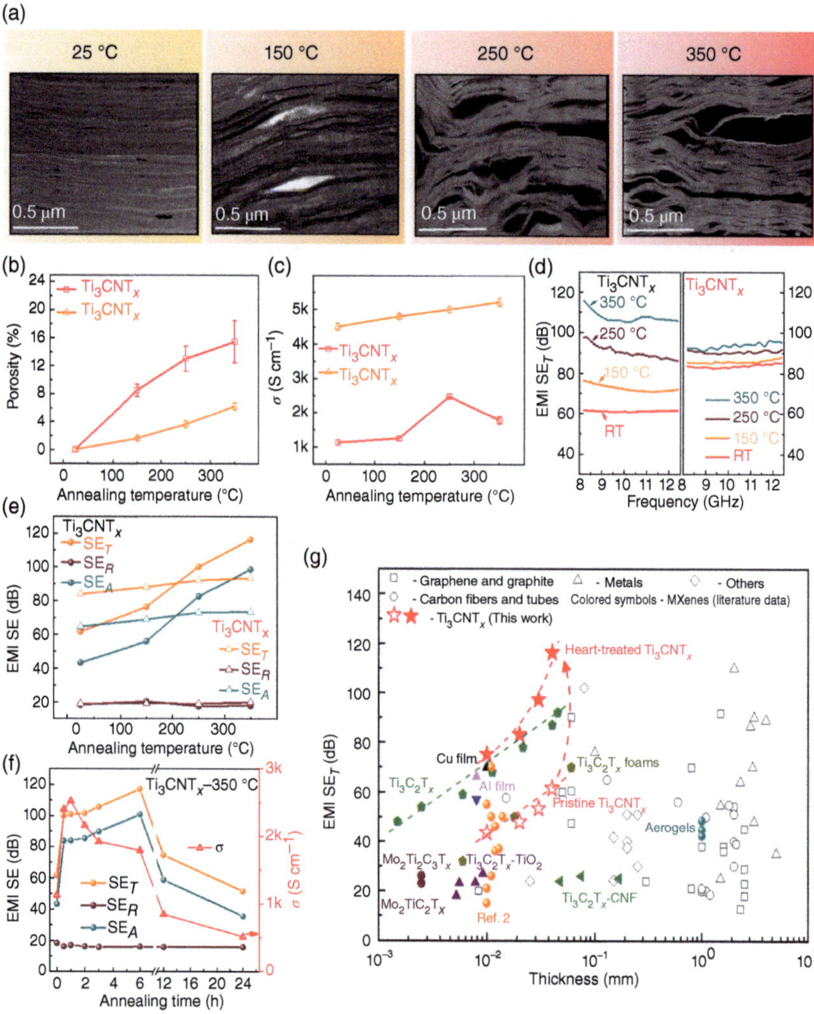

Figure 5.6 (a) Cross-sectional TEM images of Ti_3CNT_x films at different annealing temperatures. (b, c) Comparison of introduced porosity and electrical conductivity of annealed Ti_3CNT_x and $Ti_3C_2T_x$ MXenes as a function of annealing temperature. (d) EMI SE_T of 40 μm thick Ti_3CNT_x and $Ti_3C_2T_x$ films annealed at different temperatures. (e) Comparison of EMI SE_T, SE_R, and SE_A of 40 μm thick Ti_3CNT_x and $Ti_3C_2T_x$ films annealed at different temperatures. (f) EMI SE_T, SE_R, SE_A, and electrical conductivity of 40 μm thick Ti_3CNT_x films annealed at 350 °C for different times. (g) EMI SE_T as a function of the thickness for annealed Ti_3CNT_x films and materials reported in the literature. Source: Iqbal et al. [8]. Reproduced with permission of American Association for the Advancement of Science.

5.2.2 Fiber-Reinforced and Polymeric Composites of MXenes

Pure MXene laminates have shown state-of-the-art EMI shielding properties owing to their excellent metallic conductivity over a wide range (1–15 000 $S\,cm^{-1}$) and lamellar layered structure, where each layer individually acts as a shield [53, 70, 71].

However, bulk laminates suffer from weak mechanical properties owing to small 2D flakes, and the hydrophilic nature of MXenes makes them prone to oxidation, resulting in low environmental stability. Thus, significantly improved mechanical properties are required for real functional applications. Benefiting from the hydrophilicity of MXenes, MXene-based composites with reinforcing materials, including one-dimensional (1D) nanowires/nanofibers, 2D sheets of graphene or other conducting materials, and conducting/nonconducting polymers, have been developed [6, 86, 87]. In here, the higher electrical conductivity of MXenes favors the development of more conductive composites at comparatively lower MXene loadings, or addition of a nonconductive polymer marginally decreases their electrical conductivity, synergistically resulting in higher EMI SE and improved mechanical properties.

To enhance the mechanical properties of bare $Ti_3C_2T_x$ MXene, Cao et al. fabricated $Ti_3C_2T_x$–cellulose nanofiber (MXene–CNF) composite films with a nacre-like structure by facile vacuum-assisted filtration (Figure 5.7a,b) [6]. Owing to the addition of tough nanofibers, the mechanical properties of the composite films were better than those of pristine MXene and pristine CNF paper. At 50 wt% loading of d-$Ti_3C_2T_x$, the ultimate tensile strength and the fracture at strain reached 135.4 MPa and 16.7%, respectively (Figure 5.7c). In the obtained brick-and-mortar (nacre-like) structure of the d-$Ti_3C_2T_x$/CNF paper, the improved mechanical properties resulted from strong hydrogen bonding between the hydrophilic surface terminations of the MXene flakes and the hydroxyl groups of the CNFs. Owing to the hydrogen bonding, the integrated CNFs acted as stress distributors under tension, allowing the composite film to withstand higher induced strains, and the compact and well-stacked morphology of the d-$Ti_3C_2T_x$/CNF paper was retained after fracture. Moreover, the 1D CNFs decreased resistive contacts between the MXene sheets; hence, the composite films retained the maximum electrical conductivity of the $Ti_3C_2T_x$ MXene, which further increased as the MXene content increased (Figure 5.7d). A 47 μm thick nanocomposite with 80 wt% $Ti_3C_2T_x$ loading showed an electrical conductivity of 1.155 $S\,cm^{-1}$ and an EMI SE of 25.8 dB at 12.4 GHz (Figure 5.7e). Moreover, the flexible d-$Ti_3C_2T_x$/CNF paper with superior EMI shielding ability was shown to be a potential candidate for foldable and wearable electronics. In another report, commercially available hydrophilic cellulose filter paper was dip-coated with a $Ti_3C_2T_x$ MXene dispersion and then thinly coated with a polydimethylsiloxane (PDMS) solution [89]. Increasing the number of dip-coating cycles enhanced the electrical conductivity and EMI SE. In the same way, Xie et al. used aramid nanofibers (ANFs, commonly known as Kevlar) to enhance the mechanical properties of an ultrathin $Ti_3C_2T_x$–ANF composite paper [90]. A low MXene content (up to 10 wt%) ensured improved mechanical properties, with maximum ultimate tensile strength and fracture strain values of 197 MPa and 9.8%, respectively. In comparison to the d-$Ti_3C_2T_x$/CNF paper, the $Ti_3C_2T_x$–ANF composite paper exhibited much better mechanical properties and folding endurance after thousands of folding cycles. This improved performance was attributed to the ultrahigh toughness of ANF, which is the toughest synthetic material [91]. A higher content of conductive $Ti_3C_2T_x$ MXene increased the electrical conductivity

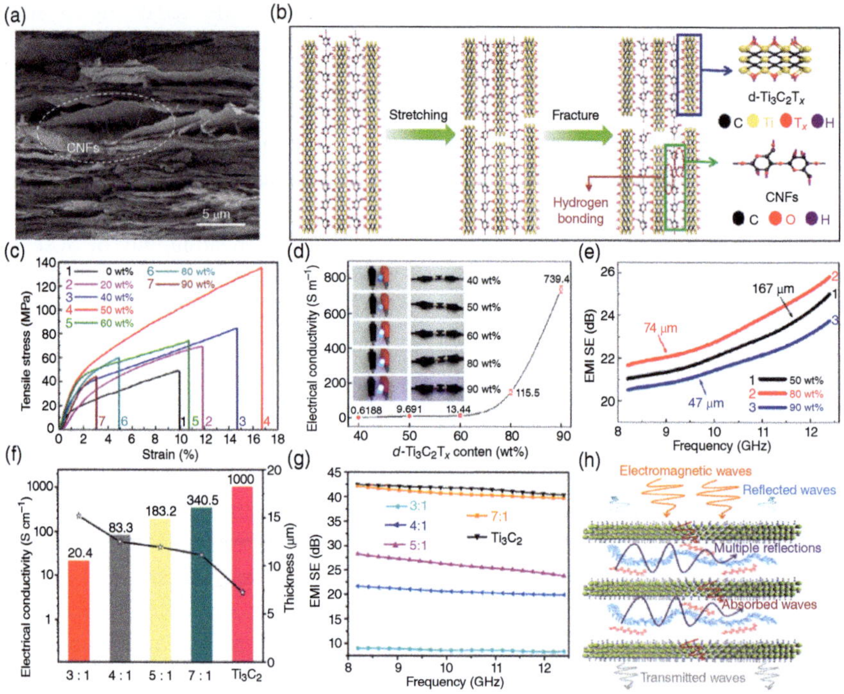

Figure 5.7 (a) SEM image of the fractured surface of d-$Ti_3C_2T_x$/CNF composite paper and (b) the fracture mechanism. (c, d) Tensile stress–strain curves and electrical conductivities of d-$Ti_3C_2T_x$/CNF composite paper sheets with varying d-$Ti_3C_2T_x$ contents. (e) EMI SE of d-$Ti_3C_2T_x$/CNF composite paper sheets with varying d-$Ti_3C_2T_x$ contents and thicknesses. Source: Cao et al. [6]. Reproduced with permission of the American Chemical Society. (f) Electrical conductivities and thicknesses and (g) EMI SE of pure $Ti_3C_2T_x$ MXene and polymeric composite films with different MXene/PANI ratios. (h) Proposed EMI shielding mechanism in the MXene/PANI composite film. Source: Liu et al. [88]. Reproduced with permission of the American Chemical Society.

of the insulating ANF paper to 170 $S\,cm^{-1}$, which resulted in an EMI SE of 33 dB at a thickness of only 17 μm.

Ma et al. developed flexible ANF/MXene/Ag nanowire (AgNW) composite films using a double-layer filtration strategy [92]. More precisely, uniformly dispersed ANFs were filtered on a nylon porous membrane to form a hydrogel. Subsequently, a homogeneous mixture of $Ti_3C_2T_x$ MXene and AgNWs at different mass ratios with respect to the ANFs (5, 10, 20, 40, 60, and 80 wt%; fixed $Ti_3C_2T_x$-to-AgNW mass ratio of 10 : 1) was filtered over the fabricated ANF layer. The obtained composite films were dried and hot pressed at 60 °C and 1 MPa, respectively. Owing to the AgNWs and the $Ti_3C_2T_x$ MXene, increasing the loading in the composite film from 20 to 80 wt% improved the electrical conductivity from 922 to 3725.6 $S\,cm^{-1}$. Similarly, the EMI SE of the composites was improved from 48.1 to 80 dB at a thickness of 91 μm.

Miao et al. developed $Ti_3C_2T_x$ MXene/AgNW composite films via a pressured extrusion method for EMI shielding [93]. In this work, CNFs were used as a binder and mixed with an aqueous dispersion of $Ti_3C_2T_x$ MXene. Dispersed AgNWs were added to the MXene/CNF mixture under stirring, and the resulting suspension with a controlled mass ratio was poured into the extruder under continuous N_2 flow at a pressure of 2 MPa. The obtained composite film containing CNFs and strong AgNWs intercalated in the MXene sheets had a tensile strength of 63.8 MPa. At a thickness of 16.9 μm, the composite film showed a maximum electrical conductivity of 3000 S cm^{-1}, with an EMI SE value of more than 42 dB in the X-band frequency range. Another study reported a $Ti_3C_2T_x$ MXene/AgNW composite film with superb flexibility and transparency for EMI shielding [94]. Liu et al. used 2D graphene oxide (GO) sheets to improve the mechanical properties of pristine $Ti_3C_2T_x$ MXene [95]. GO/MXene composite films with different GO to MXene mass ratios were fabricated by the vacuum filtration of uniformly mixed dispersions. The larger GO sheets were intercalated in the $Ti_3C_2T_x$ MXene sheets in the composite films, resulting in an improved mechanical strength of 209 MPa. The addition of GO restored the electrical conductivity of the pristine MXene, resulting in an efficient EMI SE of 50 dB at a thickness of 7 μm. Moreover, the wettability of the composite films was also improved, increasing the contact angle from 65.7° to 95.7°. This hydrophobicity makes these composite films suitable for engineering applications.

Instead of insulating 1D nanofibers, conducting polymers are also beneficial for retaining good electrical conductivity in MXene-based polymer composites. The conductive polymer chains can create conductive channels between the MXene layers, thus favoring electronic conduction and retaining the maximum electrical conductivity of the MXene sheets. In this regard, Liu et al. reported vacuum-filtered $Ti_3C_2T_x$ MXene and PEDOT:PSS (poly(sodium 4-styrene sulfonate)) nanocomposite films with superior electrical conductivity and EMI shielding properties [88]. In the brick-and-mortar-like structure, the aligned 2D MXene sheets acted as bricks, whereas the conductive chains of the organic polymer acted as mortar, and this nacre-like structure improved the mechanical properties. At a $Ti_3C_2T_x$-to-PEDOT:PSS mass ratio of 7 : 1, the electrical conductivity of pure MXene (1000 S cm^{-1}) was well maintained (340.5 S cm^{-1}) (Figure 5.7f). The stable electrical conductivity of the composite film provided an efficient EMI SE of ~42 dB in an ultrathin film of 11.1 μm, which was comparable to that of pure MXene (Figure 5.7g). Interestingly, the density of the composite film (1.94 g cm^{-3}) was much lower than that of bulk compact MXene (~2.5 g cm^{-3}). The shielding efficiency was governed by the superior electrical conductivity, layered morphology, and dipole polarization induced at the interfaces between the highly conducting $Ti_3C_2T_x$ MXene flakes and less conductive PEDOT:PSS, which attenuated the energy of the incident EMWs within the customized structure, as shown in Figure 5.7.

In another study, conducting polyaniline (PANI) was used to develop ultrathin $Ti_3C_2T_x$/PANI composite films via vacuum-assisted filtration [96]. The presence of conductive polymeric chains between the MXene layers resulted in an improved electrical conductivity of 24.4 S cm^{-1}. A 40 μm thick $Ti_3C_2T_x$/PANI composite film exhibited an EMI SE of 36 dB in the X-band frequency range.

Nonconducting polymers, if used at a controlled volume in a designed structure, are also useful for developing mechanically robust MXene composites for EMI shielding applications. For example, He et al. reported the EMI shielding of a vacuum-filtered $Ti_3C_2T_x$/hydroxyethyl cellulose (MXene/HEC) nanocomposite [87]. In the composite film structure, the HEC polymeric chains were confined between the MXene layers by hydrogen bonding, which imparted the MXene/HEC nanocomposite films with mechanical strength. The tough MXene/HEC composite film exhibited an EMI SE of 24 dB in the X-band frequency range at a thickness of 100 μm. Wang et al. annealed $Ti_3C_2T_x$ MXene at 200 °C for two hours and developed composites with an insulating epoxy using a solution-casting method [97]. The addition of epoxy significantly improved the mechanical stability of bare $Ti_3C_2T_x$ MXene. At a loading of 15 wt% annealed $Ti_3C_2T_x$ MXene, the electrical conductivity of $Ti_3C_2T_x$/epoxy composite was 1.05 S cm^{-1} as compared to 2.2×10^{-12} S cm^{-1} for the insulating epoxy. This dramatic increase in electrical conductivity enhanced the EMW attenuation ability of the composite film, resulting in an EMI SE of 41 dB at a thickness of 2 mm.

Recently, in situ polymerized dopamine was used to improve the interlayer stacking and mechanical properties of $Ti_3C_2T_x$ MXene. Owing to strong hydrogen bonding, a maximum tensile strength of 309.8 MPa was achieved, which is the highest value reported for an MXene/polymer composite. This composite also exhibited improved oxidation resistance as compared to pristine MXene. Owing to the highly compact structure and excellent electrical conductivity, the MXene/dopamine composite film with a thickness of 6.95 μm had an EMI SE of 58.4 dB in the X-band frequency range [98].

5.2.3 MXene Hybrids with Other Nanomaterials

To further enhance the intrinsic EMI SE of MXenes, hybrids have been designed with multifunctional constituents including conducting or magnetic inclusions. In hybrid systems, the introduction of multiple phases synergistically imparts mechanical strength and tunes the EMW absorption mechanism within the structure via multiple reflections by internal scattering.

Xiang et al. reported lightweight and ultrathin TiO_2–$Ti_3C_2T_x$/graphene hybrid laminate films with improved mechanical properties for EMI shielding [99]. Using vacuum filtration, a 5–9 μm thick $Ti_3C_2T_x$/GO composite film was fabricated. Subsequent annealing at 1000 °C under argon transformed this composite film into a TiO_2–$Ti_3C_2T_x$/graphene hybrid film, as the annealing temperature was above the oxidation stability limit (<800 °C) of the $Ti_3C_2T_x$ MXene sheets. Different amounts of TiO_2 were formed by the oxidation of $Ti_3C_2T_x$ MXene depending on the composition of the composite film. The addition of the highly conductive MXene decreased the surface resistance (7.5 Ω cm^{-2}) of the bulk graphene film, which then increased again after the formation of dielectric TiO_2. As reported by Han et al. the formation of TiO_2 improved the absorption of EMWs through a capacitive energy storage mechanism owing to the formation of multiple minicapacitors [100]. The moderate electrical conductivity of the TiO_2–$Ti_3C_2T_x$/graphene hybrid film with a thickness of 9.17 μm resulted in an EMI SE of 28 dB, which is suitable for commercial requirements. The absorption-dominant EMI shielding mechanism was governed by EMW energy attenuation through dielectric losses and dipolar polarization.

Similarly, Raagulan et al. reported a hydrophobic MXene- and graphene-based hybrid film with polyvinylidene fluoride (PVDF) produced by spray coating and solvent casting [101]. The electrical conductivity of 13.68 S cm^{-1} for a 0.35 mm thick MXene–graphene–PVDF hybrid film resulted in EMI SE values of 41 dB (in the S-band) and 54 dB (in the X-band) with improved thermal stability. With improved mechanical properties and a hydrophilic nature, these hybrid films have potential for use in both humid and hot environments, where other materials undergo degradation or oxidation under ambient conditions.

Li et al. reported flexible and stretchable $Ti_3C_2T_x$ MXene composites with single-walled carbon nanotubes (SWNTs) [102]. In a typical process, GO was filtered on a PVDF membrane to fabricate a 100 nm thick film, followed by the filtration of MXene/SWNT mixed dispersions. Sodium dodecyl sulfate (SDS), the surfactant used to disperse the SWNTs, was washed out with excess water. The obtained hybrid film was placed in ethanol to peel it off from the PVDF membrane and then deposited on an inflated latex balloon treated under oxygen plasma. The balloon was expanded with air/water and then deflated, resulting in crumpled MXene/SWNT composite films. The process was repeated to obtain multilayer composite films. The highly conductive SWNTs and MXene sheets resulted in films with high conductivities, leading to an EMI SE of 52 dB for a 10-layer film. This shielding performance was strain-invariant under 800% areal strains and 500 cycles of a fatigue test. Thus, this material is suitable for EMI shielding applications in flexible, stretchable, and wearable electronics.

Cao et al. developed carbon nanotube (CNT), CNF, and $Ti_3C_2T_x$ MXene hybrids that formed mechanically strong and conductive films for EMI shielding [86]. The CNTs and CNFs enhanced the mechanical properties of bare $Ti_3C_2T_x$ MXene, and a tensile strength of 97.9 MPa and fracture strain of 4.6% were achieved. Owing to the dispersed conductive CNTs within the $Ti_3C_2T_x$–CNF composite, an ultrathin hybrid film (38 μm) exhibited an efficient electrical conductivity of 25.06 S cm^{-1} and hence an EMI SE of 38.4 dB. This shielding performance was superior to that of the bare MXene–CNF nanocomposite at a comparable thickness because of the enhanced electrical conductivity, as reported in previous studies [6].

Liu et al. identified another strategy for enhancing the electrical, mechanical, and EMI shielding properties of pure MXene films [103]. The fabricated $Ti_3C_2T_x$ MXene films were immersed in different concentrations of $Al(NO_3)_3$ for 24 hours, followed by rinsing with water and drying at room temperature. Al^{3+} ions were successfully intercalated between the MXene layers, which enhanced the interfacial adhesion between the MXene flakes. Moreover, the interflake resistance was significantly reduced, resulting in improved electronic conduction between the layers. With a conductivity of 2656 S cm^{-1}, a 39 μm thick film showed an EMI SE of 80 dB in the X-band frequency range. The well-stacked MXene layers and intercalated Al^{3+} ions also improved the tensile strength from 28 MPa for the pristine MXene film to 83 MPa for the MXene/Al composite film under identical conditions.

The synergistic effect of dielectric and magnetic losses can result in more efficient EMI shielding in a broad frequency range. To observe the magnetic losses, Wang et al. used Ni chains with MXene and PVDF to produce a hybrid film by solution casting [104]. The 1D Ni chains consisted of Ni nanoparticles were embedded within MXene sheets in the MXene/Ni/PVDF hybrid films. The electrically conductive Ni chains decreased the resistance between the adjacent layers, resulting in faster electron conduction. For the P-10M10N hybrid (where M, N, and 10 represent MXene, Ni, and the loading in wt%), the electrical conductivity was 8.92 S cm^{-1} and the EMI SE value reached over 35 dB at a thickness of 0.36 mm. Similarly, Liang et al. developed hybrid films of $Ti_3C_2T_x$ MXene and Ni nanochains in wax with different MXene contents [105]. Ni chains with an average length of 20 μm and interconnected Ni nanoparticles with diameters of 300 nm were mixed with MXene sheets using a low-temperature hydrothermal process (60 °C for four hours), and paraffin-based hybrid films with a fixed Ni content (10 wt%) and varying MXene loadings were produced. At 50 wt% MXene, the electrical conductivity increased dramatically from 7.1×10^{-14} to 0.04 S cm^{-1}. The addition of MXene and Ni nanochains improved both the dielectric loss tangent ($\tan \delta_\varepsilon$) and the magnetic loss tangent ($\tan \delta_\mu$), which confirmed that the EMW absorption mechanism within the hybrid film also involved magnetic hysteresis loss. The highly conductive MXene enhanced conduction losses (from electron migration and hopping) and dielectric polarization losses (from surface terminations). An EMI SE value of ~40 dB was recorded for the hybrid film with 50 wt% MXene at a thickness of 1.3 mm. Thus, the hybrid film showed excellent potential for the enhanced absorption of EMWs under applied electromagnetic fields.

5.2.4 Layer-by-Layer (LbL) Assembly in MXene Composites

LbL assembly, which involves the repetitive stacking of two different phases, can be used to improve the EMI shielding performance and mechanical properties of MXene-based materials. Different liquid processing techniques, e.g. convenient spray coating [106], uniform spin coating [107], solution casting [108], dip coating [109], and interfacial assembly [78, 110], can be used for LbL stacking of composite films. Owing to the strong hydrophilicity MXenes, all these techniques can be used to fabricate polymeric composites of MXenes in aqueous and organic solvents [40]. Interestingly, the LbL morphology is believed to improve the mechanical strength of MXene sheets by alternating with other conductive/insulative reinforcing components.

Weng et al. fabricated semitransparent films by the alternating LbL deposition of a $Ti_3C_2T_x$–polyvinyl alcohol (PVA) dispersion and a single-/multiwalled CNT–PSS dispersion via spin-spray coating (Figure 5.8a) [111]. The LbL morphology was developed by depositing and drying one layer, followed by depositing and drying the second layer. The concentration and volume of the dispersions and the spinning speed of the substrate helped to ensure the formation of homogeneous composite films, whereas the precise thickness was controlled by the number of $Ti_3C_2T_x$–CNT bilayers. With sufficient transmission, ultrathin composite films showed promise

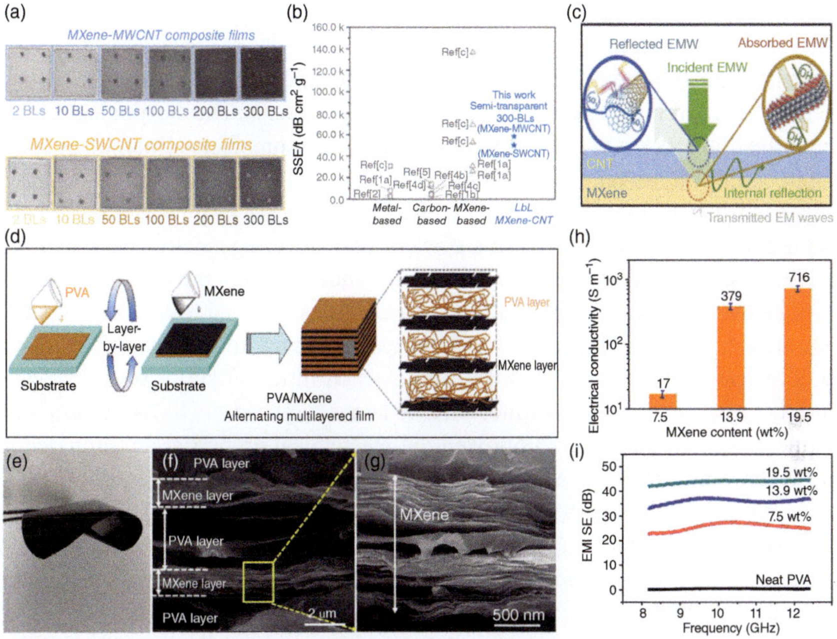

Figure 5.8 (a) LbL assembly of a $Ti_3C_2T_x$–CNT composite film. (b) Comparison of SSE/*t* for various EMI shielding materials. The $Ti_3C_2T_x$–CNT composite film shows an excellent SSE/*t* value, making it ideal for lightweight shielding materials. (c) Proposed EMI shielding mechanism of a multilayer MXene–CNT composite film. Source: Weng et al. [111]. Reproduced with permission of Wiley-VCH. (d) Schematic of the fabrication of MXene/PVA multilayer composite films via multilayer casting. (e) Digital image of the flexible MXene/PVA-AM-19.5 film. (f, g) SEM images of the MXene/PVA-AM-19.5 film at different magnifications. (h) Electrical conductivity of various PVA/MXene multilayer films. (i) EMI SE of pure PVA and various PVA/MXene multilayer films in the X-band frequency range. Source: Jin et al. [112]. Reproduced with permission of Elsevier.

for transparent electrode applications. The transmittance gradually decreased as the film thickness increased, and the electrical conductivity of a ~200 nm thick film (300 bilayer cycles) was $130\,S\,cm^{-1}$. The electrical conductivity was governed by the compact stacking of the MXene and CNTs. At the given thickness, the EMI SE and SSE/*t* values were 2.9 dB and $58\,187\,dB\,cm^2\,g^{-1}$, respectively (Figure 5.8b), because of multiple reflections in the LbL structure (Figure 5.8c). The LbL structure provided multiple interfaces between alternating layers with different impedances and enhanced multiple reflections, resulting in strong EMW absorption.

The LbL assembly of thin MXene films with polymeric materials can significantly improve the mechanical properties. Jin et al. reported dual-functional MXene/PVA composites fabricated by LbL stacking of $Ti_3C_2T_x$ MXene and PVA using an easy liquid-casting method (Figure 5.8d) [112]. An aqueous solution of PVA was cast on a clean iron substrate, followed by drying at 45 °C to completely remove the solvent. Then, an MXene dispersion with an appropriate concentration was poured onto the surface of the PVA layer and dried under the same conditions. The required thickness was obtained by repetition of these steps, ending with a PVA layer on the top covering the MXene layer. This morphology enhanced the flexibility of the composite film and lowered the oxidation kinetics of $Ti_3C_2T_x$ MXene (Figure 5.8e–g). Different numbers of cycles resulted in varying thicknesses (each MXene layer was 1–3 μm thick) and $Ti_3C_2T_x$ contents (7.5, 13.9, and 19.5 wt%). The morphology of this composite film was less compact than those of MXene composites obtained by spray or spin casting. However, with a loading of 19.5 wt% MXene, the electrical conductivity reached 7.16 S cm^{-1} (Figure 5.8h), and an EMI SE of 44.4 dB was achieved at a thickness of 25 μm (Figure 5.8i). This efficient EMI shielding resulted not only from the electrical conductivity of the composite film but also from the LbL geometry.

Unlike the uniform mixing of MXene and CNFs, as reported in the literature [6], Zhou et al. developed MXene/CNF nanocomposite films with alternating LbL stacking of MXene and CNF layers by vacuum-assisted filtration [113]. With CNF layers as the outermost layers, n inner MXene layers required $n+1$ CNF layers. The obtained multilayer structure improved the mechanical properties as compared to bare MXene and CNF films as well as the homogeneously mixed MXene/CNF films with a mechanical strength and toughness of 112.5 MPa and 2.7 MJ m^{-3}, respectively. This enhancement was governed by the crack-propagation mechanism being hindered in the multilayered structure. The alternative conducting and insulating layers also resulted in a higher EMI SE for the MXene/CNF nanocomposite film owing to enhanced internal scattering in the multilayered architecture [79]. With an electrical conductivity of 1.43 S cm^{-1}, the MXene/CNF nanocomposite film showed an EMI SE of 40 dB at a thickness of 35 μm (five CNF layers and four MXene layers). The performance of the film was well retained after 1000 folding cycles. Interestingly, this multilayered structure also showed a strong Joule heating effect, with a temperature of 100 °C obtained in only 10 seconds by applying a voltage of 6 V.

Liu et al. decorated silk fabric via the LbL deposition of $Ti_3C_2T_x$ MXene and AgNWs using vacuum-assisted spray coating [114]. With the help of atomizers, $Ti_3C_2T_x$ MXene and AgNW aqueous dispersions were repeatedly sprayed on both sides of the silk fabric and dried. The obtained leaf-like architecture, in which conductive $Ti_3C_2T_x$ flakes acted as lamina and the AgNWs as veins, enhanced the electrical conductivity of the composite, which linearly increased with the concentration of AgNWs. The addition of highly conductive metal nanowires resulted in a higher conductivity and hence enhanced the EMI SE. The conductive composite film with a thickness of 120 μm and a customized layered morphology exhibited an EMI SE of 54 dB in the X-band frequency range.

Wang et al. reported flexible and wearable textiles coated with $Ti_3C_2T_x$ MXene/polypyrrole (PPy) for EMI shielding applications [115]. The synthesized MXene sheets were modified with in situ polymerized chains of conductive PPy, and a clean polyethylene terephthalate (PET) fabric was repeatedly dipped into the dispersion to obtain the required conductive coating. Finally, a silicon coating was introduced onto the MXene/PPy-coated fabric to make the sample hydrophobic. After 10 dip-coating cycles, the electrical conductivity of the insulating PET fabric reached ~10 S cm^{-1} owing to the loading of conductive $Ti_3C_2T_x$ MXene. Stacking three pieces of the coated fabric (each with a thickness of 0.45 mm) gave an efficient EMI SE of 90 dB. The silicon coating hardly affected the electrical conductivity and EMI shielding properties but improved the lifetime of the fabric by imparting hydrophobicity. This study provided a route for synthesizing hydrophobic composites of hydrophilic MXenes for advanced real applications in severe atmospheric conditions.

Vural et al. coated a flexible PET substrate (at 70 °C) with a $Ti_3C_2T_x$ MXene/tandem repeat protein (TR42) ink [116]. Continuous printing with a resolution (minimum width of printed line) of 120 μm was obtained at a concentration of 2.25 mg ml^{-1} MXene/protein/DMSO. The printed sheet had an excellent conductivity of 1080 S cm^{-1} and an EMI SE of 50 dB at a thickness of 1.35 μm. Owing to its excellent conductivity coupled with its superior flexibility, the MXene/protein/PET composite can be used in wearable smart electronics for EMI shielding and conducting flexible electrodes.

Notably, the LbL approach widens the range of potential EMI shielding materials by reducing the skin depth effect in many conducting materials [117]. The LbL geometry creates multiple interfaces between adjacent multilayer segments and hence improves the interaction of incident EMWs within the layered morphology. In the layered structure, enhanced internal scattering from the internal interfaces and the impedance mismatch between two materials with different conductivities (or different impedances) aid in attenuating the EMW energy, resulting in a higher absorption contribution to the EMI SE.

5.2.5 Porous Structures of MXenes

For aerospace industry, military applications, and mobile electronics, efficient EMI shielding materials are qualitatively classified by weight with an ultra-lightweight being a crucial requirement. Porous architectures with extra low densities are an option for such shielding applications. A porous structure allows EMWs to enter the shield, where their energy is then dissipated through strong internal scattering. Therefore, porous materials can achieve significantly improved EMW absorption, resulting in a high EMI SE.

Liu et al. developed lightweight, flexible, and hydrophobic $Ti_3C_2T_x$ MXene foams for EMI shielding [5]. Using a typical process, compact MXene films were transformed into porous MXene foams by immersion in a hydrazine solution heated at 90 °C (Figure 5.9a). Hydrazine molecules and vapors entered the MXene layers via capillary action, separating them from each other through an expansion mechanism (Figure 5.9b–e). As a result, compact and dense films with thicknesses of 1, 3, and 6 μm were transformed into 6.2, 18, and 60 μm thick porous foams, respectively, with much lower densities. This treatment also caused the MXene film to become hydrophobic, as the hydrazine molecules replaced the –OH terminations of the MXene, generating H_2O vapor, and reduced the C–OH groups of the carbon layers, generating CO/CO_2. The removal of the oxygen-containing groups and release of these gaseous species resulted in higher Ti/O and C/O ratios in the MXene foams and cellular porous structures with water-resistant hydrophobic surfaces (Figure 5.9f,g). This approach of transforming hydrophilic pristine MXenes into hydrophobic MXenes reduced the oxidation and degradation

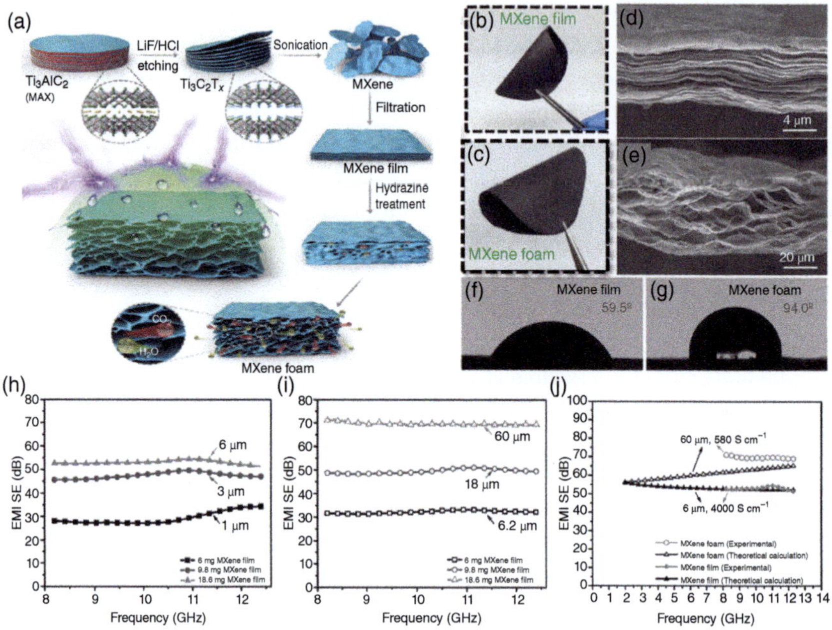

Figure 5.9 (a) Illustration of the fabrication of the hydrophobic and flexible MXene foam. Digital images of the MXene (b) film and (c) foam; cross-sectional SEM image of the MXene (d) film and (e) foam; contact angle measurements of the MXene (f) film and (g) foam; and EMI SE of the MXene (h) films and (i) foams at various thicknesses. (j) Comparison of experimental and theoretical calculations for the MXene film and foam. Source: Liu et al. [5]. Reproduced with permission of Wiley-VCH.

kinetics in humid environmental conditions. The introduction of porosity and the increased interlayer distance decreased the average electrical conductivity from 4000 S cm^{-1} (for compact films) to 588, 625, and 580 S cm^{-1}. However, despite a lower electrical conductivity, the MXene foams had higher EMI SE values than their bulk counterparts (Figure 5.9h,i). The volume expansion behavior was more evident for larger initial thicknesses, with an EMI SE of 53 dB for a 6 μm thick compact film increasing to 70 dB for a 60 μm thick foam structure. Irrespective of the lower conductivity, the higher EMI SE value was governed by the generation of multiple interfaces that acted as internal scattering sites for the incident EMWs and improved the absorption and total EMI SE of the porous MXene foams. Moreover, these results were consistent with theoretical calculations (Figure 5.9j).

Fan et al. fabricated hybrid foams of $Ti_3C_2T_x$ MXene and graphene for lightweight EMI shielding applications [118]. The dispersions with MXene to GO ratios of 1 : 1, 1 : 2, and 1 : 3 were freeze-dried at −65 °C and thermally reduced at 300 °C for one hour to obtain MXene/graphene foams with cellular porous networks. As the density of reduced graphene oxide (RGO) is lower than that of $Ti_3C_2T_x$ MXene, the addition of graphene gradually reduced the density of the hybrid foam to 7.2, 4.6, and 3.7 mg cm^{-3} (1 : 1, 1 : 2, and 1 : 3 mass loadings, respectively). Similarly, the electrical conductivity was enhanced by the addition of conductive $Ti_3C_2T_x$ MXene, with the value of 140 S cm^{-1} for bare RGO increasing to 1250 S cm^{-1} at an MXene to RGO ratio of 1 : 1. This high electrical conductivity in a porous architecture resulted in efficient EMI shielding properties. The bare RGO foam with moderate electrical conductivity showed an EMI SE of 15 dB at a thickness of 1.5 mm, whereas this value was doubled for 1 : 1 mass loading in the hybrid foam. At a larger thickness of 3 mm, the EMI SE increased to more than 50 dB for 1 : 2 mass loading in the hybrid foam. This efficient EMI shielding by the MXene/graphene hybrid foam was the result of enhanced internal scattering from multiple interfaces in the porous conductive network.

Weng et al. explored the EMI shielding mechanism of a AgNW/$Ti_3C_2T_x$ MXene hybrid foam [119]. In this study, a porous melamine formaldehyde (MF) substrate was triaxially compressed and AgNWs were grown on the buckled melamine formaldehyde (BMF) foam through repeated dip coating. The highly conductive metallic AgNWs significantly enhanced the electrical conductivity of the BMF foam. This foam was dipped into a $Ti_3C_2T_x$ MXene dispersion and freeze-dried to obtain a BMF/AgNW/MXene hybrid foam with a honeycomb structure. The electrical conductivity of the BMF/AgNW/MXene hybrid foam was further enhanced, and interestingly, it was higher than those of the bare BMF/AgNW and BMF/MXene composite foams. The highly conductive AgNWs and MXene synergistically improved the electrical conductivity of the BMF/AgNW/MXene hybrid foam and hence the EMI SE. The BMF/AgNW and BMF/MXene composite foams at a thickness of 2 mm each had an EMI SE of 40 dB, whereas the hybrid foam with a density of 49.5 mg cm^{-3} exhibited an EMI SE of 52.6 dB. These EMI SE values confirmed the important role of electrical conductivity in the EMI shielding mechanism. Absorption-dominant EMI shielding behavior was observed in the composite and hybrid foams owing to the multiple interfaces in the conductive porous structure.

Wu et al. reported the EMI shielding properties of $Ti_3C_2T_x$ MXene/SA/PDMS hybrid foams synthesized by a freeze-drying method [120]. Owing to a high electrical conductivity of 22.11 S cm^{-1}, an ultralight (20 mg cm^{-3}) MXene/SA foam with 95% MXene showed an EMI SE of 72 dB at a thickness of 2 mm. The mechanical properties of the MXene/SA foam were further improved by coating with a thin layer of PDMS via vacuum filtration. The thin layer of PDMS made the hybrid foam compressible and durable while maintaining a high EMI shielding efficiency (~50 dB), even after 500 bending and compression cycles. In contrast, an MXene/SA/PDMS hybrid fabricated by simple mixing and an MXene/SA foam with a thin PDMS coating covering the pores in the structure showed poorer EMI shielding properties (54 and 9 dB, respectively) than MXene/SA/PDMS with an interconnected porous structure, which indicated that the regular porous structure attenuated the incident EMWs more efficiently.

Similarly, Wang et al. impregnated an epoxy resin into $Ti_3C_2T_x$ MXene/carbon hybrid foams synthesized by a sol–gel method and thermal reduction [121]. Typically, catalysts such as resorcinol, formaldehyde, and sodium carbonate were added to the $Ti_3C_2T_x$ dispersion. After sonication, the mixture was cured in a glass tube at 90 °C for five hours under a continuous nitrogen flow followed by freeze-drying. The obtained porous foam was transformed into an MXene/carbon hybrid foam by thermal annealing at 400 °C for two hours. To impart mechanical strength to the hybrid foam, the porous structure was covered with an epoxy solution under vacuum. At a loading of 4.25 wt% (1.64 wt% MXene and 2.61 wt% carbon) in epoxy, the electrical conductivity of the MXene/carbon hybrid foam reached 1.84 S cm^{-1}, with an EMI SE of 46 dB in the X-band at a thickness of 2 mm. Moreover, the MXene/carbon hybrid foam exhibited a greater hardness and Young's modulus (0.31 and 3.96 GPa, respectively) than the MXene-free carbon/epoxy composite (0.28 and 3.51 GPa, respectively). This difference emphasizes the effect of conductive MXene flakes on the mechanical and EMI shielding properties of epoxy-based composites and hybrid structures.

The same group reported RGO–MXene–epoxy nanocomposites with a honeycomb structure for EMI shielding applications [122]. Using the sacrificial template method, an Al_2O_3 honeycomb substrate was used as a template for RGO deposition following $Ti_3C_2T_x$ MXene deposition by electrostatic adsorption. The mechanical properties of this architecture were reinforced by the addition of an epoxy resin. The addition of RGO and MXene synergistically improved the electrical conductivity and the resulting EMI shielding performance of the obtained structure. Interestingly, the size of the honeycomb cell was found to have an inverse effect on the electrical conductivity and EMI SE of the RGO–MXene–epoxy nanocomposite, i.e. a larger cell size resulted in a lower EMI SE and vice versa. Similarly, at a fixed RGO loading, the electrical conductivity increased as the content of conductive MXene increased. As a result, at 1.2 wt% RGO and 3.3 wt% MXene with a cell size of 0.5 mm, an electrical conductivity of 387.1 S cm^{-1} was achieved with an EMI SE of 55 dB.

Xu et al. synthesized Ti_2CT_x MXene/PVA composite foams by a freeze-drying method for EMI shielding applications [123]. Similar to the interactions observed in the multilayer films obtained by LbL assembly [112], in the Ti_2CT_x/PVA foam, the PVA molecular chains formed hydrogen bonds with the hydroxyl terminations of MXene, improving the mechanical strength and flexibility. The composite foam exhibited a lower electrical conductivity than porous $Ti_3C_2T_x$ because the electrical conductivity of Ti_2CT_x is lower than that of $Ti_3C_2T_x$. Nevertheless, the observed conductivity was far higher than those values reported for carbon-based materials [124]. This ultralight composite foam with a thickness of 5 mm exhibited an EMI SE of 26–33 dB in the X-band frequency range. At a lower density of 10.9 mg cm^{-3}, the SSE/t value reached 5136 dB cm^2 g^{-1} with only 0.15 vol% Ti_2CT_x MXene. This porous foam with a comparatively low electrical conductivity provided excellent absorption of incident EMWs with a minimal reflection contribution.

Aerogels, as apparent from the name, have much lower densities than foams but similar porous structures. The thin multiwall porous architecture helps to reduce the density of the aerogels. Bian et al. fabricated a $Ti_3C_2T_x$ MXene aerogel by conventional freeze-drying, as shown in Figure 5.10a–c, where the density was tailored by varying the concentration of the aqueous dispersions [125]. The initial concentration of the dispersion defined the resultant porous morphology owing to the concentration and size of ice crystals formed during freezing. With an ultralow density of 20.7 mg cm^{-3}, the conductive network in the $Ti_3C_2T_x$ MXene aerogel

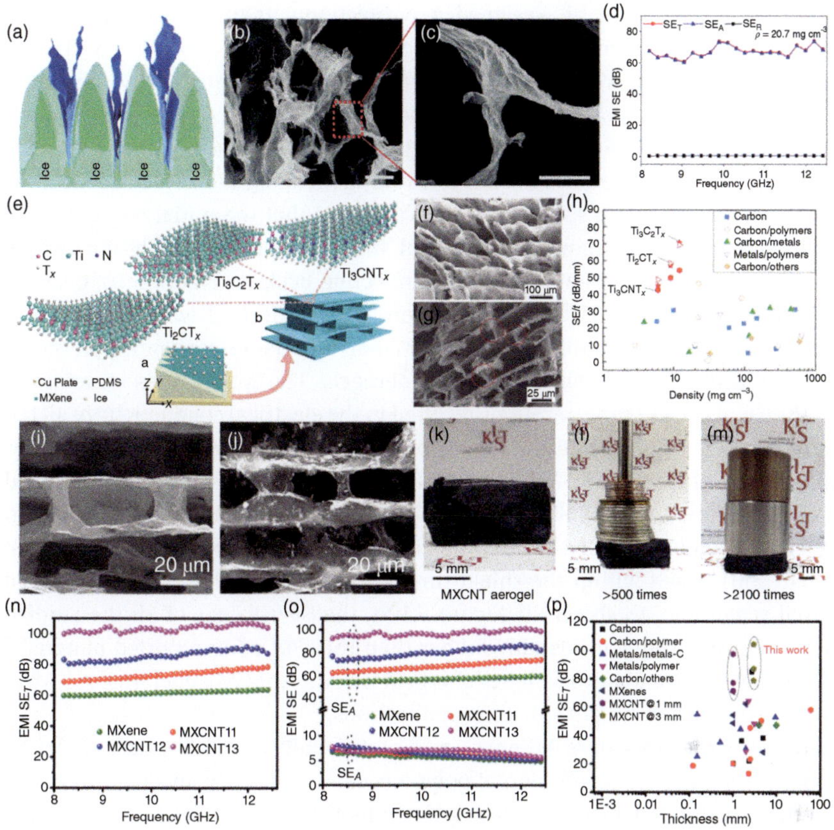

Figure 5.10 (a) Illustration of the fabrication of the $Ti_3C_2T_x$ MXene aerogel by the conventional freeze-drying method. (b, c) SEM images of the MXene aerogel at different magnifications (scale bar = 10 mm). (d) EMI SE of the MXene aerogel with a density of 20.7 mg cm^{-3}. Source: Bian et al. [125]. Reproduced with permission of the Royal Society of Chemistry. (e) Bidirectional freeze-casting mechanism. (f, g) SEM images of $Ti_3C_2T_x$ aerogels with densities of 5.5 and 11.0 mg cm^{-3}, respectively. (h) Comparison of SE/t vs. density for MXene aerogels and reported foam materials. Source: Han et al. [7]. Reproduced with permission of Wiley-VCH. (i, j) SEM images of pure $Ti_3C_2T_x$ MXene and MXCNT13 aerogels, respectively. (k–m) MXCNT13 aerogel before and during different degrees of compression. (n) SE_T and (o) SE_R and SE_A contributions for the $Ti_3C_2T_x$/CNTs aerogels. (p) Comparison of the EMI shielding performance of MXCNT aerogels (encircled) and reported porous materials. Source: Sambyal et al. [126]. Reproduced with permission of the American Chemical Society.

possessed an electrical conductivity of 22 S cm^{-1}. An EMI SE value of 75 dB was recorded at a thickness of 2 mm, whereas the SSE/t value reached 18 116 dB cm^2 g^{-1} (Figure 5.10d). These results show the potential of MXenes for lightweight EMI shielding applications, which is attributed to their excellent electrical conductivity and dimensional stability.

The MXene aerogels synthesized by the conventional freeze-drying method were fragile because of insufficient connections between adjacent MXene flakes

(Figure 5.10b,c). Han et al. designed a bidirectional freeze-drying mechanism to develop three different MXene aerogels: $Ti_3C_2T_x$, Ti_3CNT_x, and Ti_2CT_x (Figure 5.10e) [7]. In this method, a PDMS wedge was used to provide a slope for the flowing MXene dispersions of different concentrations, and a copper foil was used to simultaneously create two temperature gradients. Consequently, the formed ice crystals grew in two different directions, resulting in a long-range ordered lamellar porous architecture upon sublimation. It was found that the initial concentration of the MXene dispersion was also crucial for developing a highly conductive network consisting of MXene flakes as connecting bridges. Figure 5.10f,g shows the SEM images of the $Ti_3C_2T_x$ MXene aerogels synthesized with two different concentrations (5 and 11 mg mL^{-1}). The higher dispersion concentration led to more conductive and mechanically strong aerogels. The well-aligned structure of the MXenes at certain concentrations resulted in the electrical conductivity of all the MXene aerogels being higher than those of other carbon materials. Consequently, at a thickness of 1 mm, the $Ti_3C_2T_x$, Ti_3CNT_x, and Ti_2CT_x MXene aerogels showed absorption-dominant EMI SE values of 70.5, 69.2, and 54.1 dB, respectively, in the X-band frequency range. The porous structure maximized the attenuation of incident EMI waves within the shield through multiple interfaces and enhanced internal scattering. As shown in Figure 5.10h, the SE/t values of all the studied MXene aerogels were considerably higher than those of other reported materials, which show that conductive MXenes also provide efficient EMI shielding in porous architectures for lightweight shielding applications.

Aerogels made from pure MXenes are dimensionally stable but lack sufficient mechanical strength. The addition of polymers can improve the mechanical stability at the expense of density and electrical conductivity. In contrast, 1D reinforcing and conducting nanofibers or nanotubes can synergistically address the issues of density and electrical conductivity. Sambyal et al. established $Ti_3C_2T_x$ MXene-based aerogels with CNT inclusions (MXCNTs) using a bidirectional freeze-drying method [126]. The MXene-to-CNT ratio in the aerogels was adjusted to 1 : 1, 1 : 2, and 1 : 3. The highly conductive CNTs in the hybrid MXCNT aerogels constructed conductive channels between MXene flakes, thus maintaining the electrical conductivity of the porous network (Figure 5.10i,j). Additionally, the CNTs acted like beams to support the MXene bridges and improved the mechanical properties of the aerogel. The hybrid MXCNT aerogels exhibited greater compressive strength than bare $Ti_3C_2T_x$ MXene (Figure 5.10k–m), with the compression moduli increasing by 3898%, 7796%, and 9661% with increasing CNT loading. The maximum EMI SE value was 100 dB at an MXene to CNT weight ratio of 1 : 3 and a thickness of 3 mm, resulting in an SSE/t value of 8253.9 dB cm^2 g^{-1} (Figure 5.10n–p). Similar to other aerogels, this conductive porous structure exhibited absorption-dominant shielding behavior governed by electron migration under an applied electric field, polarization at the surface terminations, and internal scattering from multiple interfaces in the porous architecture. The hybrid $Ti_3C_2T_x$/CNT aerogels showed sufficient mechanical strength and robustness for real applications in lightweight EMI shielding materials.

Following the same approach, Zhao et al. developed $Ti_3C_2T_x$ MXene-based aerogels with conducting RGO using a freeze-drying method [127]. The hybrid

MXene/RGO aerogel showed improved dimensional stability and mechanical strength owing to the interconnected MXene and graphene flakes. Larger graphene flakes served as a core on which the smaller MXene flakes resided, thus increasing the electrical conductivity and mechanical stability of the hybrid aerogel. An electrical conductivity of 10.85 $S\,cm^{-1}$ was achieved at a very low volume fraction of MXene (0.74 vol%). As a result, the EMI SE increased to 50 dB at a thickness of 2 mm. Thus, the highly conductive CNTs resulted in better EMI shielding properties owing to more uniform mixing and dispersion throughout the porous network. The mechanism responsible for EMW attenuation involved internal scattering from multiple interfaces with large surface areas and polarization losses from surface terminations and structural defects. These MXene-based aerogels reinforced with 1D or 2D materials offer a solution for lightweight EMI shielding and absorption applications. Similarly, Wang et al. fabricated $Ti_3C_2T_x$ MXene/CNF aerogels reinforced with epoxy for mechanically strong and shapeable EMI shielding materials [128]. Zhou et al. fabricated $Ti_3C_2T_x$/calcium alginate (MXene/CA) aerogel films through Ca^{2+}-induced cross-linking [129]. The composite films were prepared by vacuum-assisted filtration and then transformed into aerogels by freeze-drying. The porosity volume of the obtained sponge-like structure was lower than that of the aerogels synthesized by unidirectional or bidirectional freeze-drying. As a result, the porous MXene/CA aerogel with an MXene loading of 90 wt% showed an EMI SE of 54.3 dB at a small thickness of 26 μm, with an SSE/t value exceeding 17 586 $dB\,cm^2\,g^{-1}$. This improved EMI shielding was caused by the porous architecture, which supported internal reflections from multiple interfaces and attenuated the EMW energy in the form of heat.

Wood-derived porous carbon (WPC) was used to fabricate ultralight aerogels with $Ti_3C_2T_x$ MXene [130]. In this study, natural wooden blocks of the required sizes were carbonized at three temperatures (500, 1000, and 1500 °C) for two hours. Higher temperature carbonization resulted in a more conductive (35.2 $S\,cm^{-1}$) and lightweight (0.17 $g\,cm^{-3}$) aerogel, which showed an average EMI SE of 57.3 dB at a thickness of 3 mm. The obtained honeycomb porous aerogel was then immersed in a highly concentrated (30 $mg\,ml^{-1}$) $Ti_3C_2T_x$ MXene dispersion. After removing the entrapped air in a vacuum oven, the MXene/WPC composites were freeze-dried for 48 hours, resulting in a building-block-like brick-and-mortar structure. The addition of MXene improved the electrical conductivity of the MXene/WPC aerogels. The synergistic effect of increased electrical conductivity and the ordered porous structure promoted EMI shielding by internal reflections; hence, the EMI SE value reached 71.3 dB under the same conditions with a slight increase in density (0.197 $g\,cm^{-3}$). The MXene/WPC aerogels also exhibited improved mechanical compression and dimensional strength along with flame-retardant and thermal insulation properties.

In summary, the porous architecture of foams and aerogels generates multiple internal interfaces. This porous structure leads to a reduction in electrical conductivity and decreases impedance mismatching, thus allowing more incident EMWs to enter the shield and significantly reducing the reflection contribution. Once the EMWs enter the porous shield, they undergo multiple reflections from the internal interfaces and are repeatedly scattered until the associated energy is lost in the form of heat. Moreover, surface functional groups and defects on MXene sheets (produced by the acidic etching conditions) help to attenuate the EMW energy through dipole polarization and the capacitive energy storage mechanism. These mechanisms make porous materials absorption-dominant shields. The excellent metallic conductivity of MXenes is also beneficial, as it remains quite high in porous foams and aerogels. The resultant efficient EMI shielding at comparatively smaller thicknesses and lower densities along with robust mechanical stability makes these porous materials suitable for real applications in military, radar, and stealth technology.

5.2.6 Segregated Structures of MXenes with Polymers

As conventional and readily available polymers such as polystyrene (PS), polyurethane, and PDMS are insulators, larger volumes of conducting fillers are needed to realize partially conducting and/or conducting composite materials. However, the use of a segregated structure can dramatically enhance the electrical conductivity of polymeric composites. In this method, a conductive filler (1D nanofibers/nanotubes or 2D nanosheets/platelets) is added in a way that develops a conductive channel of interconnected domains. For this purpose, a lower volume of conducting filler (lower percolation threshold) with a higher aspect ratio and electrical conductivity is preferred. The presence of the conducting filler in the segregated structure ensures faster electronic conduction and improved mechanical properties.

Sun et al. fabricated $Ti_3C_2T_x$ MXene@PS composites with segregated structures by an electrostatic assembly approach (Figure 5.11a) [131]. The MXene flakes, which had a zeta potential of approximately −40 mV owing to the negatively charged surface terminations, covered the positively charged PS beads of varying diameters (80, 310, and 570 μm; Figure 5.11b). The MXene sheets and PS beads were uniformly mixed, followed by vacuum-assisted filtration. Under mild compression at 130 °C for 30 minutes, highly conductive MXene@PS composite films with a very low volume fraction of MXene sheets were prepared. According to the power law, the electrical conductivity of the composite film increased dramatically at low contents of MXene because of the successful formation of a well-connected segregated structure of MXene in the composites (Figure 5.11d). At this content of MXene, the $Ti_3C_2T_x$ sheets were electrically intact to allow electronic conduction in the composite film, and the conductivity increased to 6.02 S cm^{-1} at 1.2 vol%. This enhanced electrical conductivity of the composite film at low MXene contents resulted in improved EMI shielding performance. With a further increase in the content of MXene to 1.9 vol%, the electrical conductivity reached a maximum value of 10.81 S cm^{-1} with an excellent EMI SE of more than 60 dB at a film thickness of 2 mm (Figure 5.11e).

Luo et al. developed a series of segregated conductive nanocomposites of natural rubber (NR) with $Ti_3C_2T_x$ MXene [132]. Highly flexible and stretchable films of NR with different MXene contents were fabricated by simple vacuum filtration. Electrostatic repulsion between the negatively charged MXene flakes and NR ensured uniform mixing to build a continuous interconnected network of MXene sheets in

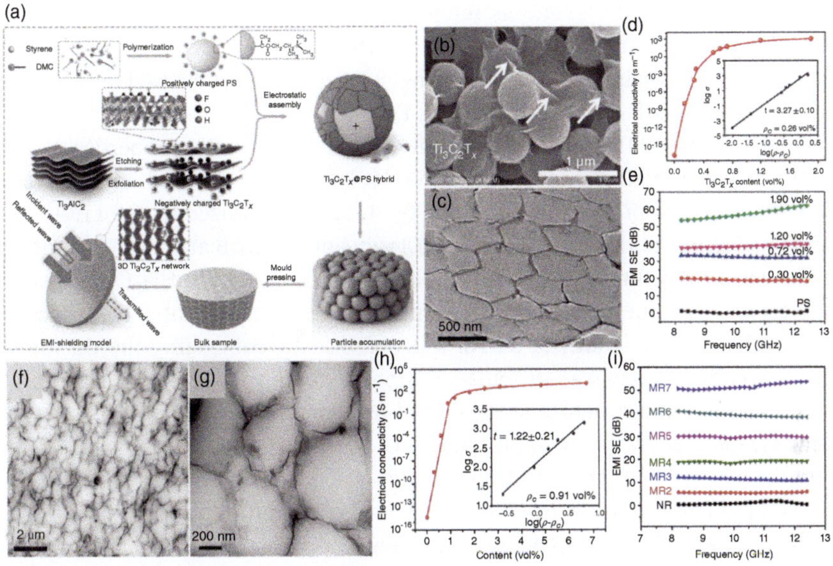

Figure 5.11 (a) Schematic of the fabrication of $Ti_3C_2T_x$@PS nanocomposites. (b, c) SEM and TEM images of the $Ti_3C_2T_x$@PS-570 hybrid with 1.20 vol% $Ti_3C_2T_x$. (d, e) Electrical conductivity and EMI SE of $Ti_3C_2T_x$@PS-570 nanocomposites. Source: Sun et al. [131]. Reproduced with permission of Wiley-VCH. (f, g) TEM images of the MXene–NR (MR5) nanocomposite. (h, i) Electrical conductivity and EMI SE of MXene–NR nanocomposites at different MXene loadings. Source: Luo et al. [132]. Reproduced with permission of Elsevier.

an insulating NR matrix (Figure 5.11f,g). The segregated structure formed in the MXene/NR nanocomposites improved the electrical conductivity of NR to 14 S cm^{-1} (Figure 5.11h). Furthermore, the EMI SE of insulating NR (transparent to EMWs with an EMI SE of 0 dB) increased linearly as the content of conductive MXene increased. As shown in Figure 5.11i, at 6.7 vol% MXene, the EMI SE value reached 53.6 dB for a 250 μm thick composite film. The formation of a conductive network helped attenuate the energy of the EMWs by conduction loss and internal scattering from multiple interfaces.

5.3 MXenes as Microwave Absorbers

Microwave absorption (MWA) is another approach for shielding electronic devices and electrical equipment from electromagnetic (EM) radiation. MWA involves the strong absorption of incident EMWs within a lossy shielding material to mitigate secondary EMI pollution by reflected EMWs. This approach is also used in radar and stealth technology applications. In the advanced technology era of electronics and warfare, microwave-absorbing materials (MWAMs) have attracted significant interest from the scientific community and defense authorities. MWAMs efficiently absorb EMWs with negligible reflection, making an object almost invisible to a radar.

The use of highly conductive MXenes as microwave absorbers is challenging. However, MXenes have a large window of electrical conductivity (1–15 000 S cm^{-1}), which provides multiple solutions for material selection. Moreover, the surface terminations, controllable surface chemistry, defects, and solution processability of MXenes as well as their ability to develop heterostructures can be used to tailor the absorption bandwidth of microwaves [133, 134].

As reported for EMI shielding, the different electromagnetic properties of the $Ti_3C_2T_x$ MXenes synthesized by the LiF + HCl and HF methods showed very different absorption properties. The $Ti_3C_2T_x$ MXene synthesized by the LiF + HCl method had a minimum reflection loss (RL) value of −27.5 dB at 55 wt% loading in paraffin wax at a thickness of 4 mm [135]. However, the HF-etched $Ti_3C_2T_x$ MXene exhibited an RL value of −42.5 dB at a much lower thickness of 1.7 mm, and a broader effective bandwidth of 5 GHz was obtained by controlling the thickness [136]. This enhancement was attributed to stronger interlayer reflections in the multilayer accordion-like structure of HF-etched MXene than in the well-delaminated MXene synthesized by the LiF + HCl method.

Han et al. studied the electromagnetic absorption properties of $Ti_3C_2T_x$ MXene-based composites in insulating wax in the X-band frequency range [100]. For this purpose, $Ti_3C_2T_x$ MXene was first annealed under argon at 800 °C, which oxidized the Ti layers to TiO_2 as the annealing temperature was above the thermal stability limit of MXene. Then, thermally annealed and oxidized MXene was mixed with the wax at different weight ratios to fabricate $Ti_3C_2T_x$–wax composites. At a loading of 50 wt% annealed MXene, the composite film showed a minimum RL value of −48.4 dB at 11.6 GHz with an effective bandwidth (the frequency range in which RL is less than −10 dB) of 2.5 GHz. The dielectric TiO_2 particles formed on the surface of the MXene sheets enhanced the polarization losses, whereas the amorphous carbon maintained the electrical conductivity of the composites.

In another similar study, Han et al. investigated the electromagnetic absorption of fully oxidized $Ti_3C_2T_x$ MXene over a wide frequency range [137]. Unlike the partial oxidation achieved in argon, $Ti_3C_2T_x$ was annealed at 800 °C in a CO_2 environment to completely transform the Ti layers into TiO_2 nanoparticles (Figure 5.12a–e). Varying amounts of the oxidized MXene were mixed with wax to produce carbon/TiO_2/wax composites. Under an applied electromagnetic field, the real and imaginary parts of the permittivity increased with increasing MXene content (Figure 5.12f,g), which was attributed to the large rutile TiO_2 particles. The composites showed a minimum RL value of −36 dB at 10.5 and 15.5 GHz for sample thicknesses of 2.2 and 1.6 mm, respectively (Figure 5.12h). The shift of the peak frequency (f_m) to a lower value with increasing sample thickness was due to the quarter-wavelength effect [138, 139]. At a thickness of 1.7 mm, the effective bandwidth covered the entire Ku-band (12.4–18 GHz). For the X-band, the effective bandwidth ranged from 8.7 to 12.3 GHz at a 2.2 mm matching thickness. The dielectric TiO_2 particles supported the formation of minicapacitors to maximize the absorption of EMW energy, as shown in Figure 5.12i.

Feng et al. comprehensively demonstrated the microwave attenuation capability of $Ti_3C_2T_x$-based MWAMs in the frequency range of 2–18 GHz [140]. At 50 wt%, the multilayer $Ti_3C_2T_x$ nanocomposites in paraffin wax exhibited improved dielectric permittivity and enhanced MWA properties with a wide effective bandwidth of

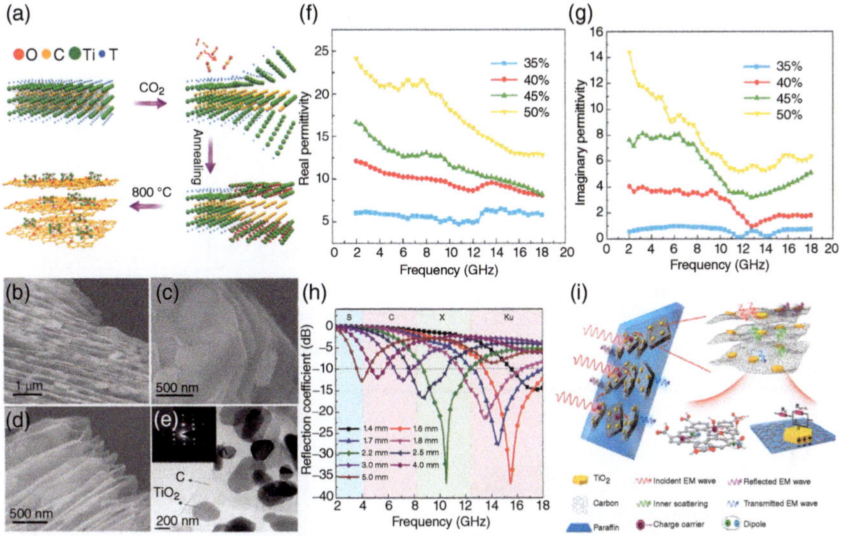

Figure 5.12 (a) Schematic of the structural evolution from $Ti_3C_2T_x$ to C/TiO_2 hybrids by heat treatment at 800 °C in a CO_2 atmosphere. (b–d) SEM images of as-synthesized C/TiO_2 hybrids derived from multilayer $Ti_3C_2T_x$ at different magnifications, showing TiO_2 particles sandwiched between electron-beam transparent carbon layers. (e) TEM image and corresponding SED pattern (inset) of a C/TiO_2 hybrid. (f) Real (ε') and (g) imaginary (ε'') permittivity vs. frequency (2–18 GHz) for composites with different loadings of C/TiO_2 hybrids in the paraffin matrix. (h) RL curves vs. frequency (2–18 GHz) and thickness (0–5 mm) for the composite with 45 wt% C/TiO_2 hybrids (S-800) in the paraffin matrix. (i) Schematic illustration of EMW absorption mechanisms for $Ti_3C_2T_x$-derived C/TiO_2 hybrids. Source: Han et al. [137]. Reproduced with permission of the American Chemical Society.

6.8 GHz (11.2–18 GHz) at a thickness of 2 mm, whereas the minimum reflection loss of −40 dB was obtained at 7.8 GHz (Figure 5.13a,b). This performance was governed by three types of electric polarization, as revealed by the Cole–Cole plots and corresponding schematic of the $Ti_3C_2T_x$/paraffin composites (Figure 5.13c). The interfacial polarization (or Maxwell–Wagner polarization) between neighboring MXene sheets with opposite charges (type I), the in-plane polarization along the direction of the stacked MXene layers (type II), and the out-of-plane polarization along the *c*-direction (type III) created microcapacitors and favored capacitive energy storage under an alternating electromagnetic field. Moreover, the strong multiple reflections within the multilayer MXene structure improved the absorption of EMWs.

The inclusion of dielectric materials introduces multiple heterogeneous interfaces, and consequently, more conductive networks are formed in the system. Tong et al. decorated a 3D multilayer RGO aerogel structure with TiO_2/$Ti_3C_2T_x$ via hydrothermal treatment [141]. The customized hybrid structure with a thickness of 2.5 mm exhibited a minimum RL value of −65.3 dB at 10 wt% loading, which is the minimum RL value obtained for MXene-based composites and/or hybrids, with an effective absorption bandwidth of 4.3 GHz. Optimized impedance matching significantly enhanced the dielectric losses in the hybrid porous structure.

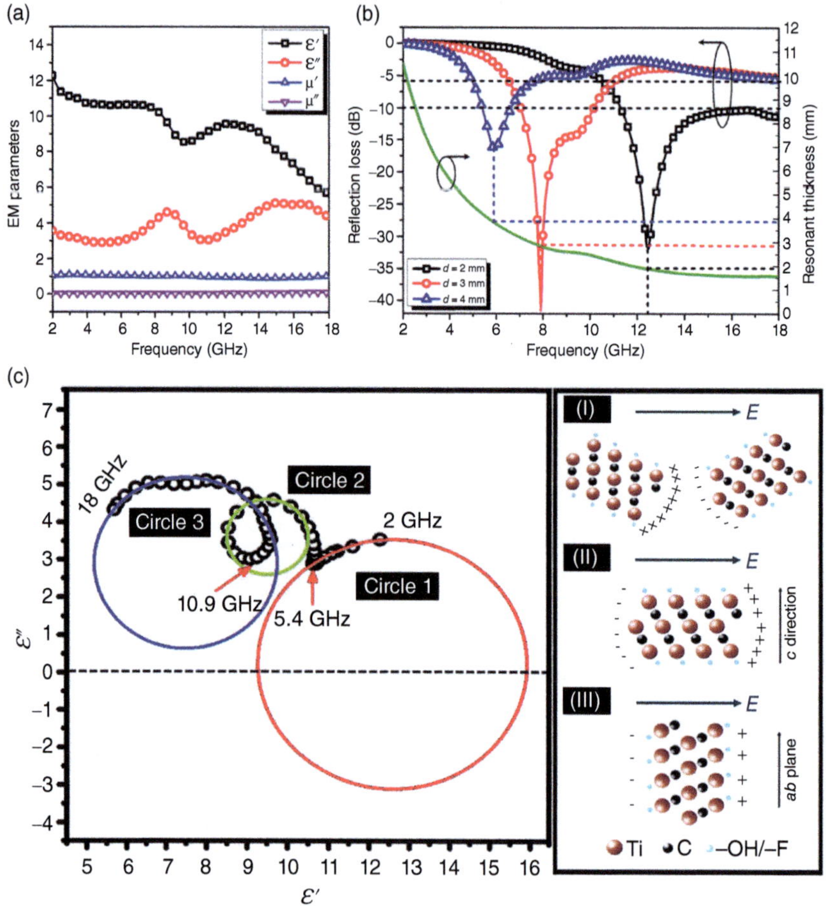

Figure 5.13 (a) Dielectric permittivity and magnetic permeability. (b) RL values of $Ti_3C_2T_x$ nanosheets/paraffin composites. (c) Typical Cole–Cole diagram over a frequency range of 2–18 GHz and the three types of electric polarization in $Ti_3C_2T_x$ composites. Source: Feng et al. [140]. Reproduced with permission of the Royal Society of Chemistry.

In another study, Ti_3AlC_2 MAX was etched in HF for different etching times (1, 3, 6, 9, and 12 hours) to obtain $Ti_3C_2T_x$ MXenes with different degrees of delamination (Figure 5.14) [142]. The interlayer distance increased at longer times owing to excessive etching. Interestingly, carbon nanospheres started to grow on the edges of the $Ti_3C_2T_x$ MXene flakes after six hours of etching owing to oxidation (Figure 5.14a–f). The size of the carbon nanospheres increased from 30 nm at 6 hours to 50 nm at 12 hours. Multilayer $Ti_3C_2T_x/TiO_2$/carbon nanosphere hybrids were fabricated and their MWA properties were analyzed. The degree of oxidation and carbon nanosphere formation enhanced the dielectric parameters and hence the RL of the $Ti_3C_2T_x/TiO_2$/carbon nanosphere hybrids (Figure 5.14g). At 50 wt% loading in wax, a minimum RL of −54.6 dB at 3.97 GHz was obtained for a 4.8 mm thick sample (Figure 5.14h). The enhanced RL was attributed to the fact that the

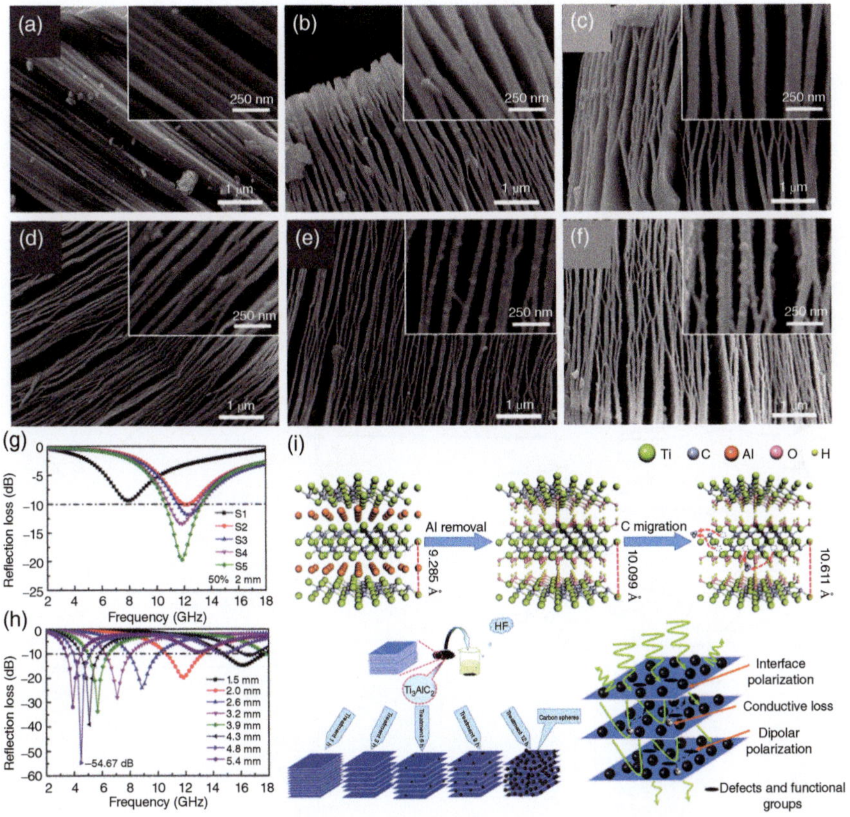

Figure 5.14 Field emission scanning electron microscopy (FESEM) images of Ti_3AlC_2 powder before and after HF treatment for different etching times: (a) Ti_3AlC_2 powder, (b) 1 hour, (c) 3 hours, (d) 6 hours, (e) 9 hours, and (f) 12 hours. The insets show corresponding high-magnification FESEM images. (g) Reflection loss curves of the S1, S2, S3, S4, and S5 paraffin composites with 50 wt% loading at a thickness of 2.0 mm. (S1–S5 indicate the increase in sample etching time.) (h) Reflection loss curves of the 50 wt% $Ti_3C_2T_x$/carbon nanosphere hybrid wax composite at various thicknesses in the frequency range of 2–18 GHz. (i) Schematics of the synthesis mechanism and microwave absorption mechanism for $Ti_3C_2T_x$/carbon nanosphere hybrids. Source: Dai et al. [142]. Reproduced with permission of the Royal Society of Chemistry.

formation of dielectric TiO_2 and carbon nanospheres enhanced the dielectric losses in addition to conduction loss in the hybrids (Figure 5.14i).

In several reports, carbon-based materials or polymers have been used to tune the impedance of conductive MXenes. $Ti_3C_2T_x$ MXene composites with CNTs grown in situ by chemical vapor deposition (CVD) were fabricated for EMW absorption applications (Figure 5.15a) [143]. Uniformly grown 1D CNTs on the 2D MXene sheets (Figure 5.15b,c) resulted in a hierarchical sandwich structure, which improved the real and imaginary permittivity (Figure 5.15d,e) and hence the RL as shown in Figure 5.15f, which reached a minimum value of −52.9 dB (Figure 5.15h) in the range of 2–18 GHz at 35 wt% loading in wax. The composites showed an

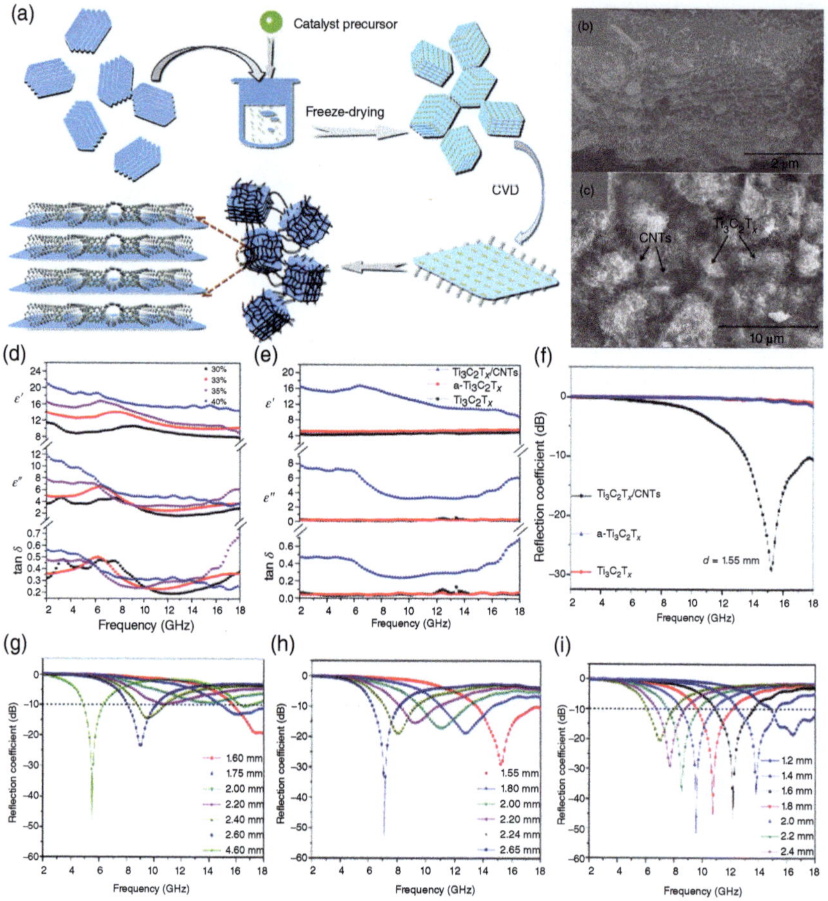

Figure 5.15 (a) Schematic of the synthesis of $Ti_3C_2T_x$/CNT hybrids by CVD. (b, c) SEM images of the $Ti_3C_2T_x$/CNT composite at different magnifications. (d) Real (ε') and imaginary (ε'') parts of the permittivity and tangent loss (tan δ) of $Ti_3C_2T_x$/CNT with different loadings. (e) Permittivity (real (ε') and imaginary (ε'')) and tangent loss (tan δ) as a function of frequency for $Ti_3C_2T_x$, a-$Ti_3C_2T_x$, and $Ti_3C_2T_x$/CNT composites in wax at 35 wt% loading. (f) Reflection coefficients of $Ti_3C_2T_x$, a-$Ti_3C_2T_x$, and $Ti_3C_2T_x$/CNT composites in wax at a thickness of 1.55 mm. (g–i) Theoretical reflection coefficient (RC) curves of $Ti_3C_2T_x$/CNT at various thicknesses with filler loadings of 30, 35, and 40 wt%. Source: Li et al. [143]. Reproduced with permission of the Royal Society of Chemistry.

effective absorption bandwidth of 4.46 GHz at a thickness of 1.55 mm, which was higher than those of pristine and annealed multilayer MXenes. Increasing the thickness to 5 mm further enhanced the bandwidth to 14.54 GHz (3.46–18 GHz; Figure 5.15g–i).

Intrinsically conducting polymers can also be used in MXene-based composites for MWA. Using oxidative polymerization, Wei et al. prepared in situ grown PANI composites with $Ti_3C_2T_x$ MXene [144]. The sandwich structure of the obtained MXene/PANI composites, in which the PANI chains grew within the multilayer MXene sheets, showed improved absorption behavior that was tunable depending on the PANI content. For the 1.8 mm thick sample in paraffin, the minimum RL value of −56.30 dB was obtained at a frequency of 13.8 GHz. In the thickness range of 1.8–2.6 mm, the effective absorption bandwidth covered the entire X-band and Ku-band regions, showing promise for MWA applications. The lower conductivity of PANI aided in impedance matching to maximize the waves entering the multilayered structure, which were then attenuated through multiple reflections.

Liu et al. used PPy microspheres with $Ti_3C_2T_x$ MXene to fabricate heterostructured composites with core–shell structures [145]. At a low loading of 10 wt% in wax and a thickness of 3.6 mm, the MXene/PPy composite exhibited a minimum RL value of −49.5 dB at 7.6 GHz with an effective absorption bandwidth of 5.14 dB (8.55–15.18 GHz). The effective absorption bandwidth could be widened by controlling the thickness and the MXene/PPy loading. Similarly, Tong et al. decorated multilayer $Ti_3C_2T_x$ MXene structures with PPy chains through in situ oxidative polymerization [146]. The heterostructured composite with 25 wt% MXene/PPy and a thickness of 3.2 mm showed an RL value of −49.2 dB at 8.5 GHz. The maximum effective absorption bandwidth of 13.7 GHz (4.3–18 GHz) was achieved in the thickness range of 1.5–5 mm.

The synergistic effect of highly conductive $Ti_3C_2T_x$ MXene and less conductive polymers enhanced impedance matching and thus promoted multiple conductive paths and dielectric losses by interfacial polarization and dipolar polarization. The MXene structure obtained by HF etching was found to be more effective for MWA owing to increased multiple reflections in its multilayered morphology.

The single dielectric loss mechanism limits the extent to which the absorption behavior of conducting materials can be enhanced. Thus, to further enhance the MWA properties of MXenes, a synergistic improvement of dielectric and magnetic losses is crucial. The addition of magnetic inclusions can improve the magnetic permeability of a system, hence introducing magnetic loss and widening the absorption bandwidth. He et al. used HF-etched $Ti_3C_2T_x$ MXene to develop composites with magnetic FeCo nanoparticles by in situ hydrothermal synthesis [147]. The minimum RL value of 17.86 dB was obtained at a thickness of 1.6 mm. The magnetic nanoparticles on the surface and within the MXene sheets enhanced the permeability of the composite; hence, a broader absorption bandwidth of 8.8 GHz was achieved at a thickness of 1.6 mm.

In another study, Fe_3O_4 nanoparticles were grown on the surface and in the interlayer space of $Ti_3C_2T_x$ MXene using a simple solvothermal process (Figure 5.16a) [148]. Unlike the hydrothermal process, solvothermal synthesis provided a reducing environment, which protected the MXene sheets from oxidation at 200 °C. As shown in Figure 5.16b, the $Fe_3O_4@Ti_3C_2T_x$ composites exhibited ferromagnetic behavior, where the saturation magnetization (M_s) linearly increased with the Fe_3O_4 loading. The inclusion of magnetic particles slightly decreased the dielectric permittivity of MXene (Figure 5.16c,d), whereas the magnetic permeability was improved (Figure 5.16e,f). As a result, a 4.2 mm thick $Fe_3O_4@Ti_3C_2T_x$ composite with 25 wt% loading in wax exhibited a minimum RL value of −57.2 dB at 15.7 GHz with an absorption bandwidth of 1.4 GHz (Figure 5.16g), which was attributed to magnetic loss as well as strong dielectric polarization loss (Figure 5.16h).

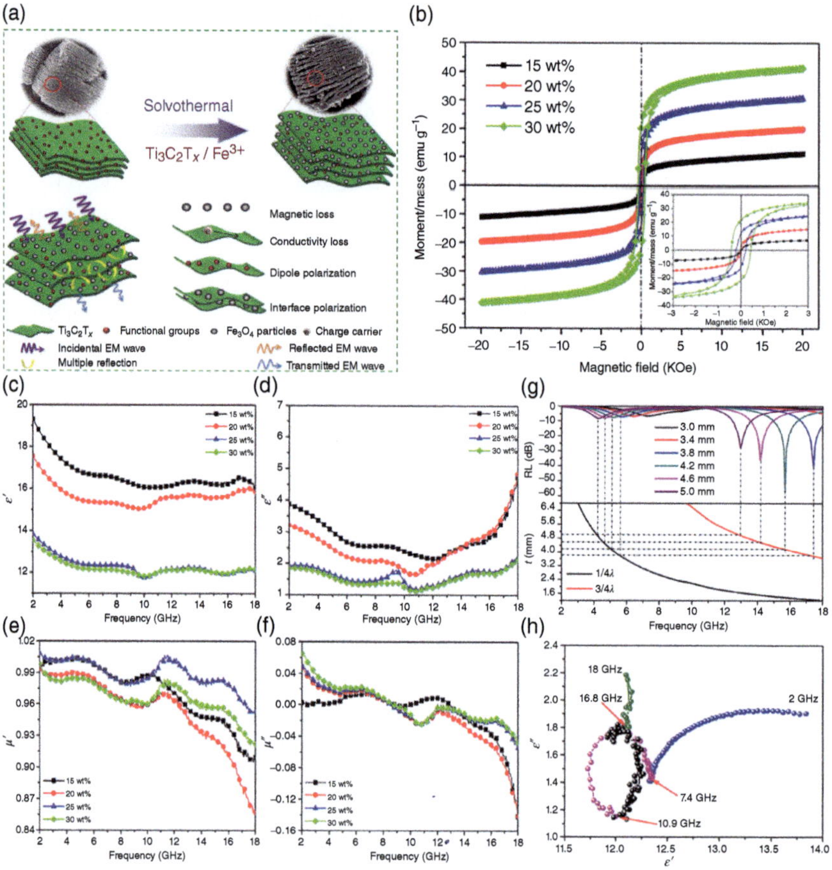

Figure 5.16 (a) Schematic of the synthesis and MWA mechanisms of $Fe_3O_4@Ti_3C_2T_x$ composites. (b) Magnetic hysteresis loops of $Fe_3O_4@Ti_3C_2T_x$ composites at various Fe_3O_4 contents. (c–f) Frequency-dependent dielectric and magnetic parameters. (g) Frequency-dependent RL curves of the 25 wt% $Fe_3O_4@Ti_3C_2T_x$ composite at various thicknesses and simulations of the absorber thickness (t_m) as a function of peak frequency (f_m) using the quarter-wavelength model. (h) Cole–Cole plot for the 25 wt% $Fe_3O_4@Ti_3C_2T_x$ composite. Source: Zhang et al. [148]. Reproduced with permission of Elsevier.

Deng et al. fabricated Co_3O_4-decorated $Ti_3C_2T_x$ composites by hydrothermal synthesis using a strong base under an inert atmosphere followed by thermal annealing [149]. At 50 wt% loading in wax, the $Ti_3C_2T_x/Co_3O_4$ composite with a thickness of 2 mm showed an RL value of −55.4 dB with an effective absorption broadband of 6.2 GHz (10.8–17 GHz). Yang et al. developed layered polyvinyl butyral (PVB)/$Ba_3Co_2Fe_{24}O_{41}/Ti_3C_2T_x$ MXene composites using a simple tape-casting method [150]. At 30 wt% loading and a thickness of 2.8 mm, a minimum RL value of −46.3 dB was achieved at 5.8 GHz with an effective absorption bandwidth of 1.6 GHz. Liang et al. reported $Ti_3C_2T_x$ MXene/Ni chain composites for both EMI shielding and MWA, which exhibited a minimum RL value of −49.9 dB at 11.9 GHz at a thickness of 1.75 mm [105]. The effective absorption bandwidth could be increased to 2.1 GHz by tuning the thickness and Ni/MXene content.

The use of ferrites is a simple approach for improving the MWA properties of MXenes because their temperate permeability and high specific resistance can reduce the skin depth effect. Li et al. utilized magnetic Ni–Zn ferrite ($Ni_{0.5}Zn_{0.5}Fe_2O_4$) particles to develop $Ti_3C_2T_x$-based nanocomposites for minimum reflection loss [151]. Ferrites were used to enhance magnetic losses along with the dielectric losses occurring in the conductive MXene. At 5 wt% $Ti_3C_2T_x$, the Ni–Zn ferrite/MXene composite exhibited a minimum RL value of −42.5 dB at 13.5 GHz with a broad effective bandwidth of ~3 GHz (12–15 GHz). This excellent MWA performance was attributed to dielectric loss, magnetic loss, conductivity loss, and internal scattering from multiple interfaces. Similarly, Liu et al. developed $Ti_3C_2T_x$ MXene composites with $Co_{0.2}Ni_{0.2}Zn_{0.4}Fe_2O_4$ (CNZF) ferrite using hydrothermal synthesis [152]. The dielectric and magnetic losses were enhanced in the synthesized MXene/ferrite composites, which exhibited a minimum RL of −58.4 dB at 6.3 GHz. The absorption bandwidth reached 2.2 GHz (3.8–6.0 GHz) at a thickness of 4.2 mm.

The synergistic effect of dielectric losses caused by conducting $Ti_3C_2T_x$ MXene and magnetic losses governed by magnetic inclusions improve the MWA properties by reducing the skin depth effect and maximizing polarization and hysteresis losses. This combination results in the broadband absorption of EMWs in multilayer MXene composites.

5.4 Summary

The emergence of highly conductive MXenes has inspired considerable advancements in the field of EMI shielding materials. Since the first report in 2016, MXenes have become the most studied material for EMI shielding and absorption. Owing to their outstanding metallic conductivity, robust flexibility, and excellent mechanical properties, MXenes are promising for EMI shielding over a thickness range of a few nanometers, which is crucial for advanced 5G electronics and portable devices. The shielding efficiency further increases with the thickness of MXene laminates and composites, whereas hybrids and heterostructures exhibit excellent mechanical strength for real applications. MXene foams and aerogels exhibit efficient EMI shielding and high SSE/t values at much lower weights, making them ideal for lightweight aerospace applications. In segregated structures, the high aspect ratio and excellent electrical conductivity of MXenes result in efficient and cost-effective EMI shielding materials. For microwave absorption, different structural designs and chemical compositions of MXenes have resulted in minimum RL values over broad frequency ranges at very small thicknesses. Therefore, MXene-based materials in various structural forms meet the challenges of EMI shielding in advanced technologies. As research on MXenes is still emerging, many aspects of their shielding properties remain unexplored. As the EMI shielding efficiency can be further improved by tailoring magnetic losses, MXenes with intrinsic magnetic properties should be investigated. Overall, MXenes are the leading EMI shielding materials for a wide variety of applications, including smarter and faster portable electronics, medical appliances, telecommunication devices, military equipment, and radar and stealth technology.

References

1 Xu, B. and Gogotsi, Y. (2020). MXenes – the fastest growing materials family in the two-dimensional world. *Chinese Chemical Letters* 31 (4): 919–921.

2 Kumar, P., Yu, S., Shahzad, F. et al. (2016). Ultrahigh electrically and thermally conductive self-aligned graphene/polymer composites using large-area reduced graphene oxides. *Carbon* 101: 120–128.

3 Kumar, P., Shahzad, F., Yu, S. et al. (2015). Large-area reduced graphene oxide thin film with excellent thermal conductivity and electromagnetic interference shielding effectiveness. *Carbon* 94: 494–500.

4 Shahzad, F., Alhabeb, M., Hatter, C.B. et al. (2016). Electromagnetic interference shielding with 2D transition metal carbides (MXenes). *Science* 353 (6304): 1137–1140.

5 Liu, J., Zhang, H.B., Sun, R. et al. (2017). Hydrophobic, flexible, and lightweight MXene foams for high-performance electromagnetic-interference shielding. *Advanced Materials* 29 (38): 1702367.

6 Cao, W.T., Chen, F.F., Zhu, Y.J. et al. (2018). Binary strengthening and toughening of MXene/cellulose nanofiber composite paper with nacre-inspired structure and superior electromagnetic interference shielding properties. *ACS Nano* 12 (5): 4583–4593.

7 Han, M., Yin, X., Hantanasirisakul, K. et al. (2019). Anisotropic MXene aerogels with a mechanically tunable ratio of electromagnetic wave reflection to absorption. *Advanced Optical Materials* 7 (10): 1900267.

8 Iqbal, A., Shahzad, F., Hantanasirisakul, K. et al. (2020). Anomalous absorption of electromagnetic waves by 2D transition metal carbonitride Ti_3CNT_x (MXene). *Science* 369 (6502): 446–450.

9 Lei, J.-C., Zhang, X., and Zhou, Z. (2015). Recent advances in MXene: preparation, properties, and applications. *Frontiers of Physics* 10 (3): 276–286.

10 Zhu, J., Ha, E., Zhao, G. et al. (2017). Recent advance in MXenes: a promising 2D material for catalysis, sensor and chemical adsorption. *Coordination Chemistry Reviews* 352: 306–327.

11 Hantanasirisakul, K., Zhao, M.Q., Urbankowski, P. et al. (2016). Fabrication of Ti_3C_2Tx MXene transparent thin films with tunable optoelectronic properties. *Advanced Electronic Materials* 2 (6): 1600050.

12 Kim, S.J., Koh, H.-J., Ren, C.E. et al. (2018). Metallic $Ti_3C_2T_x$ MXene gas sensors with ultrahigh signal-to-noise ratio. *ACS Nano* 12 (2): 986–993.

13 Kong, W., Niu, Y., Liu, M. et al. (2019). One-step hydrothermal synthesis of fluorescent MXene-like titanium carbonitride quantum dots. *Inorganic Chemistry Communications* 105: 151–157.

14 Guan, Q., Ma, J., Yang, W. et al. (2019). Highly fluorescent Ti_3C_2 MXene quantum dots for macrophage labeling and Cu^{2+} ion sensing. *Nanoscale* 11 (30): 14123–14133.

15 Park, T.H., Yu, S., Koo, M. et al. (2019). Shape-adaptable 2D titanium carbide (MXene) heater. *ACS Nano* 13 (6): 6835–6844.

16 Tu, S., Jiang, Q., Zhang, X. et al. (2018). Large dielectric constant enhancement in MXene percolative polymer composites. *ACS Nano* 12 (4): 3369–3377.

17 Dong, Y., Mallineni, S.S.K., Maleski, K. et al. (2018). Metallic MXenes: a new family of materials for flexible triboelectric nanogenerators. *Nano Energy* 44: 103–110.

18 Kim, H., Anasori, B., Gogotsi, Y. et al. (2017). Thermoelectric properties of two-dimensional molybdenum-based MXenes. *Chemistry of Materials* 29 (15): 6472–6479.

19 Malaki, M., Maleki, A., and Varma, R.S. (2019). MXenes and ultrasonication. *Journal of Materials Chemistry A* 7 (18): 10843–10857.

20 Naguib, M., Mochalin, V.N., Barsoum, M.W. et al. (2014). 25th anniversary article: MXenes: a new family of two-dimensional materials. *Advanced Materials* 26 (7): 992–1005.

21 Naguib, M., Kurtoglu, M., Presser, V. et al. (2012). Two-dimensional transition metal carbides. *ACS Nano* 6 (2): 1322–1331.

22 Naguib, M., Kurtoglu, M., Presser, V. et al. (2011). Two-dimensional nanocrystals produced by exfoliation of Ti_3AlC_2. *Advanced Materials* 23 (37): 4248–4253.

23 Alhabeb, M., Maleski, K., Mathis, T.S. et al. (2018). Selective etching of silicon from Ti_3SiC_2 (MAX) to obtain 2D titanium carbide (MXene). *Angewandte Chemie International Edition* 57 (19): 5444–5448.

24 Nowotny, V.H. (1971). Strukturchemie einiger verbindungen der übergangsmetalle mit den elementen C, Si, Ge, Sn. *Progress in Solid State Chemistry* 5: 27–70.

25 Barsoum, M.W. and El-Raghy, T. (1996). Synthesis and characterization of a remarkable ceramic: Ti_3SiC_2. *Journal of the American Ceramic Society* 79 (7): 1953–1956.

26 Anasori, B., Lukatskaya, M.R., and Gogotsi, Y. (2017). 2D metal carbides and nitrides (MXenes) for energy storage. *Nature Reviews Materials* 2 (2): 1–17.

27 Sokol, M., Natu, V., Kota, S. et al. (2019). On the chemical diversity of the MAX phases. *Trends in Chemistry* 1 (2): 210–223.

28 Barsoum, M.W. (2000). The $M_{N+1}AX_N$ phases: a new class of solids: thermodynamically stable nanolaminates. *Progress in Solid State Chemistry* 28 (1–4): 201–281.

29 Khazaei, M., Ranjbar, A., Arai, M. et al. (2017). Electronic properties and applications of MXenes: a theoretical review. *Journal of Materials Chemistry C* 5 (10): 2488–2503.

30 Hantanasirisakul, K. and Gogotsi, Y. (2018). Electronic and optical properties of 2D transition metal carbides and nitrides (MXenes). *Advanced Materials* 30 (52): 1804779.

31 Anasori, B., Xie, Y., Beidaghi, M. et al. (2015). Two-dimensional, ordered, double transition metals carbides (MXenes). *ACS Nano* 9 (10): 9507–9516.

32 Rosen, J., Dahlqvist, M., Tao, Q. et al. (2019). In-and out-of-plane ordered MAX phases and their MXene derivatives. In: *2D Metal Carbides and Nitrides (MXenes)* (eds. B. Anasori and Y. Gogotsi), 37–52. Springer.

33 Novoselov, K.S., Geim, A.K., Morozov, S.V. et al. (2004). Electric field effect in atomically thin carbon films. *Science* 306 (5696): 666–669.

34 Koenig, S.P., Doganov, R.A., Schmidt, H. et al. (2014). Electric field effect in ultrathin black phosphorus. *Applied Physics Letters* 104 (10): 103106.

35 Alhabeb, M., Maleski, K., Anasori, B. et al. (2017). Guidelines for synthesis and processing of two-dimensional titanium carbide ($Ti_3C_2T_x$ MXene). *Chemistry of Materials* 29 (18): 7633–7644.

36 Hope, M.A., Forse, A.C., Griffith, K.J. et al. (2016). NMR reveals the surface functionalisation of Ti_3C_2 MXene. *Physical Chemistry Chemical Physics* 18 (7): 5099–5102.

37 Hu, M., Hu, T., Li, Z. et al. (2018). Surface functional groups and interlayer water determine the electrochemical capacitance of $Ti_3C_2T_x$ MXene. *ACS Nano* 12 (4): 3578–3586.

38 Naguib, M., Halim, J., Lu, J. et al. (2013). New two-dimensional niobium and vanadium carbides as promising materials for Li-ion batteries. *Journal of the American Chemical Society* 135 (43): 15966–15969.

39 Mashtalir, O., Naguib, M., Mochalin, V.N. et al. (2013). Intercalation and delamination of layered carbides and carbonitrides. *Nature Communications* 4: 1716.

40 Maleski, K., Mochalin, V.N., and Gogotsi, Y. (2017). Dispersions of two-dimensional titanium carbide MXene in organic solvents. *Chemistry of Materials* 29 (4): 1632–1640.

41 Naguib, M., Unocic, R.R., Armstrong, B.L. et al. (2015). Large-scale delamination of multi-layers transition metal carbides and carbonitrides "MXenes". *Dalton Transactions* 44 (20): 9353–9358.

42 Overbury, S.H., Kolesnikov, A.I., Brown, G.M. et al. (2018). Complexity of intercalation in MXenes: destabilization of urea by two-dimensional titanium carbide. *Journal of the American Chemical Society* 140 (32): 10305–10314.

43 Shen, L., Zhou, X., Zhang, X. et al. (2018). Carbon-intercalated $Ti_3C_2T_x$ MXene for high-performance electrochemical energy storage. *Journal of Materials Chemistry A* 6 (46): 23513–23520.

44 Lin, H., Wang, X., Yu, L. et al. (2017). Two-dimensional ultrathin MXene ceramic nanosheets for photothermal conversion. *Nano Letters* 17 (1): 384–391.

45 Ghidiu, M., Lukatskaya, M.R., Zhao, M.Q. et al. (2014). Conductive two-dimensional titanium carbide 'clay' with high volumetric capacitance. *Nature* 516 (7529): 78–U171.

46 Lipatov, A., Alhabeb, M., Lukatskaya, M.R. et al. (2016). Effect of synthesis on quality, electronic properties and environmental stability of individual monolayer Ti_3C_2 MXene flakes. *Advanced Electronic Materials* 2 (12): 1600255.

47 Lukatskaya, M.R., Mashtalir, O., Ren, C.E. et al. (2013). Cation intercalation and high volumetric capacitance of two-dimensional titanium carbide. *Science* 341 (6153): 1502–1505.

48 Levitt, A.S., Alhabeb, M., Hatter, C.B. et al. (2019). Electrospun MXene/carbon nanofibers as supercapacitor electrodes. *Journal of Materials Chemistry A* 7 (1): 269–277.

49 Wang, Z., Qin, S., Seyedin, S. et al. (2018). High-performance biscrolled MXene/carbon nanotube yarn supercapacitors. *Small* 14 (37): 1802225.

50 Boota, M., Anasori, B., Voigt, C. et al. (2016). Pseudocapacitive electrodes produced by oxidant-free polymerization of pyrrole between the layers of 2D titanium carbide (MXene). *Advanced Materials* 28 (7): 1517–1522.

51 Fu, Q., Wang, X., Zhang, N. et al. (2018). Self-assembled $Ti_3C_2T_x$/SCNT composite electrode with improved electrochemical performance for supercapacitor. *Journal of Colloid and Interface Science* 511: 128–134.

52 Wang, L., Zhang, H., Wang, B. et al. (2016). Synthesis and electrochemical performance of $Ti_3C_2T_x$ with hydrothermal process. *Electronic Materials Letters* 12 (5): 702–710.

53 Han, M., Shuck, C.E., Rakhmanov, R. et al. (2020). Beyond $Ti_3C_2T_x$: MXenes for electromagnetic interference shielding. *ACS Nano* 14 (4): 5008–5016.

54 Zhao, C., Wang, Q., Zhang, H. et al. (2016). Two-dimensional titanium carbide/RGO composite for high-performance supercapacitors. *ACS Applied Materials & Interfaces* 8 (24): 15661–15667.

55 Shan, Q., Mu, X., Alhabeb, M. et al. (2018). Two-dimensional vanadium carbide (V_2C) MXene as electrode for supercapacitors with aqueous electrolytes. *Electrochemistry Communications* 96: 103–107.

56 Mashtalir, O., Lukatskaya, M.R., Zhao, M.-Q. et al. (2015). Amine-assisted delamination of Nb_2C MXene for Li-ion energy storage devices. *Advanced Materials* 27 (23): 3501–3506.

57 Halim, J., Kota, S., Lukatskaya, M.R. et al. (2016). Synthesis and characterization of 2D molybdenum carbide (MXene). *Advanced Functional Materials* 26 (18): 3118–3127.

58 Tao, Q., Dahlqvist, M., Lu, J. et al. (2017). Two-dimensional $Mo_{1.33}C$ MXene with divacancy ordering prepared from parent 3D laminate with in-plane chemical ordering. *Nature Communications* 8 (1): 14949.

59 Halim, J., Lukatskaya, M.R., Cook, K.M. et al. (2014). Transparent conductive two-dimensional titanium carbide epitaxial thin films. *Chemistry of Materials* 26 (7): 2374–2381.

60 Liu, F., Zhou, A., Chen, J. et al. (2017). Preparation of Ti_3C_2 and Ti_2C MXenes by fluoride salts etching and methane adsorptive properties. *Applied Surface Science* 416: 781–789.

61 Feng, A., Yu, Y., Jiang, F. et al. (2017). Fabrication and thermal stability of NH_4HF_2-etched Ti_3C_2 MXene. *Ceramics International* 43 (8): 6322–6328.

62 Yang, S., Zhang, P., Wang, F. et al. (2018). Fluoride-free synthesis of two-dimensional titanium carbide (MXene) using a binary aqueous system. *Angewandte Chemie* 130 (47): 15717–15721.

63 Peng, C., Wei, P., Chen, X. et al. (2018). A hydrothermal etching route to synthesis of 2D MXene (Ti_3C_2, Nb_2C): enhanced exfoliation and improved adsorption performance. *Ceramics International* 44 (15): 18886–18893.

64 He, P., Wang, X.-X., Cai, Y.-Z. et al. (2019). Tailoring $Ti_3C_2T_x$ nanosheets to tune local conductive network as an environmentally friendly material for highly efficient electromagnetic interference shielding. *Nanoscale* 11 (13): 6080–6088.

65 Shuck, C.E., Han, M., Maleski, K. et al. (2019). Effect of Ti_3AlC_2 MAX phase on structure and properties of resultant $Ti_3C_2T_x$ MXene. *ACS Applied Nano Materials* 2 (6): 3368–3376.

66 Berdiyorov, G. (2015). Effect of surface functionalization on the electronic transport properties of Ti_3C_2 MXene. *EPL (Europhysics Letters)* 111 (6): 67002.

67 Chae, Y., Kim, S.J., Cho, S.-Y. et al. (2019). An investigation into the factors governing the oxidation of two-dimensional Ti_3C_2 MXene. *Nanoscale* 11 (17): 8387–8393.

68 Zhang, J., Kong, N., Hegh, D. et al. (2020). Freezing titanium carbide (MXenes) aqueous dispersions for ultra-long-term storage. *ACS Applied Materials & Interfaces* 12 (30): 34032–34040.

69 Sang, X., Xie, Y., Lin, M.-W. et al. (2016). Atomic defects in monolayer titanium carbide ($Ti_3C_2T_x$) MXene. *ACS Nano* 10 (10): 9193–9200.

70 Zhang, C., Anasori, B., Seral-Ascaso, A. et al. (2017). Transparent, flexible, and conductive 2D titanium carbide (MXene) films with high volumetric capacitance. *Advanced Materials* 29 (36): 1702678.

71 Zhang, J., Kong, N., Uzun, S. et al. (2020). Scalable manufacturing of free-standing, strong $Ti_3C_2T_x$ MXene films with outstanding conductivity. *Advanced Materials* 32: 2001093.

72 Kim, D., Ko, T.Y., Kim, H. et al. (2019). Nonpolar organic dispersion of 2D $Ti_3C_2T_x$ MXene flakes via simultaneous interfacial chemical grafting and phase transfer method. *ACS Nano* 13 (12): 13818–13828.

73 Zhao, S., Li, L., Zhang, H.-B. et al. (2020). Janus MXene nanosheets for macroscopic assemblies. *Materials Chemistry Frontiers* 4 (3): 910–917.

74 Caffrey, N.M. (2018). Effect of mixed surface terminations on the structural and electrochemical properties of two-dimensional $Ti_3C_2T_2$ and V_2CT_2 MXenes multilayers. *Nanoscale* 10 (28): 13520–13530.

75 Schultz, T., Frey, N.C., Hantanasirisakul, K. et al. (2019). Surface termination dependent work function and electronic properties of $Ti_3C_2T_x$ MXene. *Chemistry of Materials* 31 (17): 6590–6597.

76 Hart, J.L., Hantanasirisakul, K., Lang, A.C. et al. (2019). Control of MXenes' electronic properties through termination and intercalation. *Nature Communications* 10.

77 Wei, Q., Pei, S., Qian, X. et al. Superhigh electromagnetic interference shielding of ultrathin aligned pristine graphene nanosheets film. *Advanced Materials*: 1907411.

78 Yun, T., Kim, H., Iqbal, A. et al. (2020). Electromagnetic interference shielding: electromagnetic shielding of monolayer MXene assemblies (Adv. Mater. 9/2020). *Advanced Materials* 32 (9): 2070064.

79 Iqbal, A., Sambyal, P., and Koo, C.M. 2D MXenes for electromagnetic shielding: a review. *Advanced Functional Materials*: 2000883.

80 Li, Z., Wang, Z., Lu, W. et al. (2018). Theoretical study of electromagnetic interference shielding of 2D MXenes films. *Metals* 8 (8): 652.

81 Simon, R.M. (1981). EMI shielding through conductive plastics. *Polymer-Plastics Technology and Engineering* 17 (1): 1–10.

82 Moore, R. (2016). *Electromagnetic Composites Handbook*, 2e. McGraw-Hill Education.

83 Born, M., Wolf, E., and Bhatia, A.B. (2016). *Principles of Optics Electromagnetic Theory of Propagation, Interference and Diffraction of Light*. Cambridge: Cambridge University Press.

84 Katsidis, C.C. and Siapkas, D.I. (2002). General transfer-matrix method for optical multilayer systems with coherent, partially coherent, and incoherent interference. *Applied Optics* 41 (19): 3978–3987.

85 Li, X., Yin, X., Liang, S. et al. (2019). 2D carbide MXene Ti_2CT_X as a novel high-performance electromagnetic interference shielding material. *Carbon* 146: 210–217.

86 Cao, W., Ma, C., Tan, S. et al. (2019). Ultrathin and flexible CNTs/MXene/cellulose nanofibrils composite paper for electromagnetic interference shielding. *Nano-Micro Letters* 11 (1): 72.

87 He, P., Cao, M.-S., Cai, Y.-Z. et al. (2020). Self-assembling flexible 2D carbide MXene film with tunable integrated electron migration and group relaxation toward energy storage and green EMI shielding. *Carbon* 157: 80–89.

88 Liu, R., Miao, M., Li, Y. et al. (2018). Ultrathin biomimetic polymeric $Ti_3C_2T_x$ MXene composite films for electromagnetic interference shielding. *ACS Applied Materials & Interfaces* 10 (51): 44787–44795.

89 Hu, D., Huang, X., Li, S. et al. (2020). Flexible and durable cellulose/MXene nanocomposite paper for efficient electromagnetic interference shielding. *Composites Science and Technology* 188: 107995.

90 Xie, F., Jia, F., Zhuo, L. et al. (2019). Ultrathin MXene/aramid nanofiber composite paper with excellent mechanical properties for efficient electromagnetic interference shielding. *Nanoscale* 11 (48): 23382–23391.

91 Yang, M., Cao, K., Sui, L. et al. (2011). Dispersions of aramid nanofibers: a new nanoscale building block. *ACS Nano* 5 (9): 6945–6954.

92 Ma, Z., Kang, S., Ma, J. et al. (2020). Ultraflexible and mechanically strong double-layered aramid nanofiber–$Ti_3C_2T_x$ MXene/silver nanowire nanocomposite papers for high-performance electromagnetic interference shielding. *ACS Nano* 14 (7): 8368–8382.

93 Miao, M., Liu, R., Thaiboonrod, S. et al. (2020). Silver nanowires intercalating $Ti_3C_2T_x$ MXene composite films with excellent flexibility for electromagnetic interference shielding. *Journal of Materials Chemistry C* 8 (9): 3120–3126.

94 Chen, W., Liu, L.-X., Zhang, H.-B. et al. (2020). Flexible, transparent, and conductive $Ti_3C_2T_x$ MXene–silver nanowire films with smart acoustic sensitivity for high-performance electromagnetic interference shielding. *ACS Nano* 14 (12): 16643–16653.

95 Liu, J., Liu, Z., Zhang, H.-B. et al. (2020). Ultrastrong and highly conductive MXene-based films for high-performance electromagnetic interference shielding. *Advanced Electronic Materials* 6 (1): 1901094.

96 Zhang, Y., Wang, L., Zhang, J. et al. (2019). Fabrication and investigation on the ultra-thin and flexible $Ti_3C_2T_x$/co-doped polyaniline electromagnetic interference shielding composite films. *Composites Science and Technology* 183: 107833.

97 Wang, L., Chen, L., Song, P. et al. (2019). Fabrication on the annealed $Ti_3C_2T_x$ MXene/epoxy nanocomposites for electromagnetic interference shielding application. *Composites Part B: Engineering* 171: 111–118.

98 Lee, G.S., Yun, T., Kim, H. et al. (2020). Mussel inspired highly aligned $Ti_3C_2T_x$ MXene film with synergistic enhancement of mechanical strength and ambient stability. *ACS Nano* 14 (9): 11722–11732.

99 Xiang, C., Guo, R., Lin, S. et al. (2019). Lightweight and ultrathin TiO_2-$Ti_3C_2T_X$/graphene film with electromagnetic interference shielding. *Chemical Engineering Journal* 360: 1158–1166.

100 Han, M., Yin, X., Wu, H. et al. (2016). Ti_3C_2 MXenes with modified surface for high-performance electromagnetic absorption and shielding in the X-band. *ACS Applied Materials & Interfaces* 8 (32): 21011–21019.

101 Raagulan, K., Braveenth, R., Jang, H.J. et al. (2018). Electromagnetic shielding by MXene-graphene-PVDF composite with hydrophobic, lightweight and flexible graphene coated fabric. *Materials* 11 (10): 1803.

102 Li, Y., Tian, X., Gao, S.-P. et al. (2020). Reversible crumpling of 2D titanium carbide (MXene) nanocoatings for stretchable electromagnetic shielding and wearable wireless communication. *Advanced Functional Materials* 30 (5): 1907451.

103 Liu, Z., Zhang, Y., Zhang, H.-B. et al. (2020). Electrically conductive aluminum ion-reinforced MXene films for efficient electromagnetic interference shielding. *Journal of Materials Chemistry C* 8 (5): 1673–1678.

104 Wang, S.-J., Li, D.-S., and Jiang, L. (2019). Synergistic effects between MXenes and Ni chains in flexible and ultrathin electromagnetic interference shielding films. *Advanced Materials Interfaces* 6 (19): 1900961.

105 Liang, L., Han, G., Li, Y. et al. (2019). Promising $Ti_3C_2T_x$ MXene/Ni chain hybrid with excellent electromagnetic wave absorption and shielding capacity. *ACS Applied Materials & Interfaces* 11 (28): 25399–25409.

106 Carey, T., Jones, C., Le Moal, F. et al. (2018). Spray-coating thin films on three-dimensional surfaces for a semitransparent capacitive-touch device. *ACS Applied Materials & Interfaces* 10 (23): 19948–19956.

107 Jo, J.W., Jung, J.W., Lee, J.U. et al. (2010). Fabrication of highly conductive and transparent thin films from single-walled carbon nanotubes using a new non-ionic surfactant via spin coating. *ACS Nano* 4 (9): 5382–5388.

108 Meitl, M.A., Zhou, Y., Gaur, A. et al. (2004). Solution casting and transfer printing single-walled carbon nanotube films. *Nano Letters* 4 (9): 1643–1647.

109 Sundaram, H.S., Han, X., Nowinski, A.K. et al. (2014). One-step dip coating of zwitterionic sulfobetaine polymers on hydrophobic and hydrophilic surfaces. *ACS Applied Materials & Interfaces* 6 (9): 6664–6671.

110 Yun, T., Kim, J.-S., Shim, J. et al. (2017). Ultrafast interfacial self-assembly of 2D transition metal dichalcogenides monolayer films and their vertical and in-plane heterostructures. *ACS Applied Materials & Interfaces* 9 (1): 1021–1028.

111 Weng, G.-M., Li, J., Alhabeb, M. et al. (2018). Layer-by-layer assembly of cross-functional semi-transparent MXene-carbon nanotubes composite films for next-generation electromagnetic interference shielding. *Advanced Functional Materials* 28 (44): 1803360.

112 Jin, X., Wang, J., Dai, L. et al. (2020). Flame-retardant poly(vinyl alcohol)/MXene multilayered films with outstanding electromagnetic interference shielding and thermal conductive performances. *Chemical Engineering Journal* 380: 122475.

113 Zhou, B., Zhang, Z., Li, Y. et al. (2020). Flexible, robust, and multifunctional electromagnetic interference shielding film with alternating cellulose nanofiber and MXene layers. *ACS Applied Materials & Interfaces* 12 (4): 4895–4905.

114 Liu, L.-X., Chen, W., Zhang, H.-B. et al. (2019). Flexible and multifunctional silk textiles with biomimetic leaf-like MXene/silver nanowire nanostructures for electromagnetic interference shielding, humidity monitoring, and self-derived hydrophobicity. *Advanced Functional Materials* 29 (44): 1905197.

115 Wang, Q.-W., Zhang, H.-B., Liu, J. et al. (2019). Multifunctional and water-resistant MXene-decorated polyester textiles with outstanding electromagnetic interference shielding and joule heating performances. *Advanced Functional Materials* 29 (7): 1806819.

116 Vural, M., Pena-Francesch, A., Bars-Pomes, J. et al. (2018). Inkjet printing of self-assembled 2D titanium carbide and protein electrodes for stimuli-responsive electromagnetic shielding. *Advanced Functional Materials* 28 (32): 1801972.

117 Ramírez-Herrera, C.A., Gonzalez, H., de la Torre, F. et al. (2019). Electrical properties and electromagnetic interference shielding effectiveness of interlayered systems composed by carbon nanotube filled carbon nanofiber mats and polymer composites. *Nanomaterials* 9 (2): 238.

118 Fan, Z., Wang, D., Yuan, Y. et al. (2020). A lightweight and conductive MXene/graphene hybrid foam for superior electromagnetic interference shielding. *Chemical Engineering Journal* 381: 122696.

119 Weng, C., Wang, G., Dai, Z. et al. (2019). Buckled AgNW/MXene hybrid hierarchical sponges for high-performance electromagnetic interference shielding. *Nanoscale* 11 (47): 22804–22812.

120 Wu, X., Han, B., Zhang, H.-B. et al. (2020). Compressible, durable and conductive polydimethylsiloxane-coated MXene foams for high-performance electromagnetic interference shielding. *Chemical Engineering Journal* 381: 122622.

121 Wang, L., Qiu, H., Song, P. et al. (2019). 3D $Ti_3C_2T_x$ MXene/C hybrid foam/epoxy nanocomposites with superior electromagnetic interference shielding performances and robust mechanical properties. *Composites Part A: Applied Science and Manufacturing* 123: 293–300.

122 Song, P., Qiu, H., Wang, L. et al. (2020). Honeycomb structural rGO-MXene/epoxy nanocomposites for superior electromagnetic interference shielding performance. *Sustainable Materials and Technologies* 24: e00153.

123 Xu, H., Yin, X., Li, X. et al. (2019). Lightweight Ti_2CT_x MXene/poly(vinyl alcohol) composite foams for electromagnetic wave shielding with absorption-dominated feature. *ACS Applied Materials & Interfaces* 11 (10): 10198–10207.

124 Li, X.-H., Li, X., Liao, K.-N. et al. (2016). Thermally annealed anisotropic graphene aerogels and their electrically conductive epoxy composites with excellent electromagnetic interference shielding efficiencies. *ACS Applied Materials & Interfaces* 8 (48): 33230–33239.

125 Bian, R., He, G., Zhi, W. et al. (2019). Ultralight MXene-based aerogels with high electromagnetic interference shielding performance. *Journal of Materials Chemistry C* 7 (3): 474–478.

126 Sambyal, P., Iqbal, A., Hong, J. et al. (2019). Ultralight and mechanically robust $Ti_3C_2T_x$ hybrid aerogel reinforced by carbon nanotubes for electromagnetic interference shielding. *ACS Applied Materials & Interfaces* 11 (41): 38046–38054.

127 Zhao, S., Zhang, H.B., Luo, J.Q. et al. (2018). Highly electrically conductive three-dimensional $Ti_3C_2T_X$ MXene/reduced graphene oxide hybrid aerogels with excellent electromagnetic interference shielding performances. *ACS Nano* 12 (11): 11193–11202.

128 Wang, L., Song, P., Lin, C.-T. et al. (2020). 3D shapeable, superior electrically conductive cellulose nanofibers/$Ti_3C_2T_x$ MXene aerogels/epoxy nanocomposites for promising EMI shielding. *Research* 2020: 4093732.

129 Zhou, Z., Liu, J., Zhang, X. et al. (2019). Ultrathin MXene/calcium alginate aerogel film for high-performance electromagnetic interference shielding. *Advanced Materials Interfaces* 6 (6): 1802040.

130 Liang, C., Qiu, H., Song, P. et al. (2020). Ultra-light MXene aerogel/wood-derived porous carbon composites with wall-like "mortar/brick" structures for electromagnetic interference shielding. *Science Bulletin* 65 (8): 616–622.

131 Sun, R., Zhang, H.B., Liu, J. et al. (2017). Highly conductive transition metal carbide/carbonitride (MXene)@ polystyrene nanocomposites fabricated by electrostatic assembly for highly efficient electromagnetic interference shielding. *Advanced Functional Materials* 27 (45).

132 Luo, J.-Q., Zhao, S., Zhang, H.-B. et al. (2019). Flexible, stretchable and electrically conductive MXene/natural rubber nanocomposite films for efficient electromagnetic interference shielding. *Composites Science and Technology* 182: 107754.

133 Li, M., Han, M., Zhou, J. et al. (2018). Novel scale-like structures of graphite/TiC/Ti_3C_2 hybrids for electromagnetic absorption. *Advanced Electronic Materials* 4 (5): 1700617.

134 Luo, H., Feng, W., Liao, C. et al. (2018). Peaked dielectric responses in Ti_3C_2 MXene nanosheets enabled composites with efficient microwave absorption. *Journal of Applied Physics* 123 (10): 104103.

135 Cui, G., Sun, X., Zhang, G. et al. (2019). Electromagnetic absorption performance of two-dimensional MXene $Ti_3C_2T_x$ exfoliated by HCl+ LiF etchant with diverse etching times. *Materials Letters* 252: 8–10.

136 Tong, Y., He, M., Zhou, Y. et al. (2018). Electromagnetic wave absorption properties in the centimetre-band of $Ti_3C_2T_x$ MXenes with diverse etching time. *Journal of Materials Science: Materials in Electronics* 29 (10): 8078–8088.

137 Han, M., Yin, X., Li, X. et al. (2017). Laminated and two-dimensional carbon-supported microwave absorbers derived from MXenes. *ACS Applied Materials & Interfaces* 9 (23): 20038–20045.

138 Liu, X., Guo, H., Xie, Q. et al. (2015). Enhanced microwave absorption properties in GHz range of Fe_3O_4/C composite materials. *Journal of Alloys and Compounds* 649: 537–543.

139 Ma, Z., Zhang, Y., Cao, C. et al. (2011). Attractive microwave absorption and the impedance match effect in zinc oxide and carbonyl iron composite. *Physica B: Condensed Matter* 406 (24): 4620–4624.

140 Feng, W., Luo, H., Wang, Y. et al. (2018). Ti_3C_2 MXene: a promising microwave absorbing material. *RSC Advances* 8 (5): 2398–2403.

141 Tong, Y., He, M., Zhou, Y. et al. (2018). Three-dimensional hierarchical architecture of the TiO_2/$Ti_3C_2T_x$/RGO ternary composite aerogel for enhanced electromagnetic wave absorption. *ACS Sustainable Chemistry & Engineering* 6 (7): 8212–8222.

142 Dai, B., Zhao, B., Xie, X. et al. (2018). Novel two-dimensional $Ti_3C_2T_x$ MXenes/nano-carbon sphere hybrids for high-performance microwave absorption. *Journal of Materials Chemistry C* 6 (21): 5690–5697.

143 Li, X., Yin, X., Han, M. et al. (2017). Ti_3C_2 MXenes modified with in situ grown carbon nanotubes for enhanced electromagnetic wave absorption properties. *Journal of Materials Chemistry C* 5 (16): 4068–4074.

144 Wei, H., Dong, J., Fang, X. et al. (2019). $Ti_3C_2T_x$ MXene/polyaniline (PANI) sandwich intercalation structure composites constructed for microwave absorption. *Composites Science and Technology* 169: 52–59.

145 Liu, T., Liu, N., An, Q. et al. (2019). Designed construction of $Ti_3C_2T_x$@ PPY composites with enhanced microwave absorption performance. *Journal of Alloys and Compounds* 802: 445–457.

146 Tong, Y., He, M., Zhou, Y. et al. (2018). Hybridizing polypyrrole chains with laminated and two-dimensional $Ti_3C_2T_x$ toward high-performance electromagnetic wave absorption. *Applied Surface Science* 434: 283–293.

147 He, J., Shan, D., Yan, S. et al. (2019). Magnetic FeCo nanoparticles-decorated Ti_3C_2 MXene with enhanced microwave absorption performance. *Journal of Magnetism and Magnetic Materials* 492: 165639.

148 Zhang, X., Wang, H., Hu, R. et al. (2019). Novel solvothermal preparation and enhanced microwave absorption properties of $Ti_3C_2T_x$ MXene modified by in situ coated Fe_3O_4 nanoparticles. *Applied Surface Science* 484: 383–391.

149 Deng, R., Chen, B., Li, H. et al. (2019). MXene/Co_3O_4 composite material: stable synthesis and its enhanced broadband microwave absorption. *Applied Surface Science* 488: 921–930.

150 Yang, H., Dai, J., Liu, X. et al. (2017). Layered PVB/$Ba_3Co_2Fe_{24}O_{41}$/Ti_3C_2 Mxene composite: enhanced electromagnetic wave absorption properties with high impedance match in a wide frequency range. *Materials Chemistry and Physics* 200: 179–186.

151 Li, Y., Zhou, X., Wang, J. et al. (2017). Facile preparation of in situ coated $Ti_3C_2T_x$/$Ni_{0.5}Zn_{0.5}Fe_2O_4$ composites and their electromagnetic performance. *RSC Advances* 7 (40): 24698–24708.

152 Liu, P., Yao, Z., Ng, V.M.H. et al. (2019). Novel multilayer-like structure of $Ti_3C_2T_x$/CNZF composites for low-frequency electromagnetic absorption. *Materials Letters* 248: 214–217.

6

Other 2D Materials

6.1 Introduction

As described in Chapters 4 and 5, graphene and transition metal carbides/carbonitrides (MXenes) are at the forefront of electromagnetic interference (EMI) shielding applications because these two-dimensional (2D) materials have high electrical conductivities and outstanding EMI shielding properties. However, modern highly integrated devices also require superior microwave absorption materials to avoid secondary electromagnetic pollution. In the past decade, other 2D nanomaterials have been explored for EMI shielding and microwave absorption, such as transition metal dichalcogenides (TMDCs, e.g. MoS_2, WS_2, and TaS_2) [1–4], hexagonal boron nitride (h-BN) [5], black phosphorus (BP), and metal–organic frameworks (MOFs) [6, 7]. These 2D materials possess moderate or tunable electrical conductivities, 2D sheet structures, high surface areas, and excellent mechanical stabilities, making them promising candidates for electromagnetic wave (EMW) attenuation. This chapter highlights the EMI shielding performance of these 2D materials. This compiled information will provide an understanding of the future challenges and requirements for designing ideal 2D shielding materials for next-generation electronic systems.

6.2 2D Materials Beyond Graphene and MXenes

6.2.1 Molybdenum Disulfide (MoS_2)

MoS_2 belongs to the TMDC family. Bulk MoS_2 has a sandwich structure; it consists of a monolayer of Mo atoms sandwiched between two layers of S atoms with an interlayer spacing of 0.62 nm and a free spacing of 0.30 nm as shown in Figure 6.1a [9]. Bulk MoS_2, which is a black powder, is chemically stabilized by the saturated S atoms (except the edges) present on the basal plane of individual MoS_2 monolayers [1]. These MoS_2 monolayers stack together through different atomic stacking configurations to yield one of two crystal structures (triangular prism 2H phase or octahedral 1T phase) (Figure 6.1b,c) [1]. In bulk MoS_2, there are strong covalent Mo—S bonds within the monolayer sheets and weak van der Waals forces between

Two-Dimensional Materials for Electromagnetic Shielding, First Edition.
Chong Min Koo, Pradeep Sambyal, Aamir Iqbal, Faisal Shahzad, and Junpyo Hong.

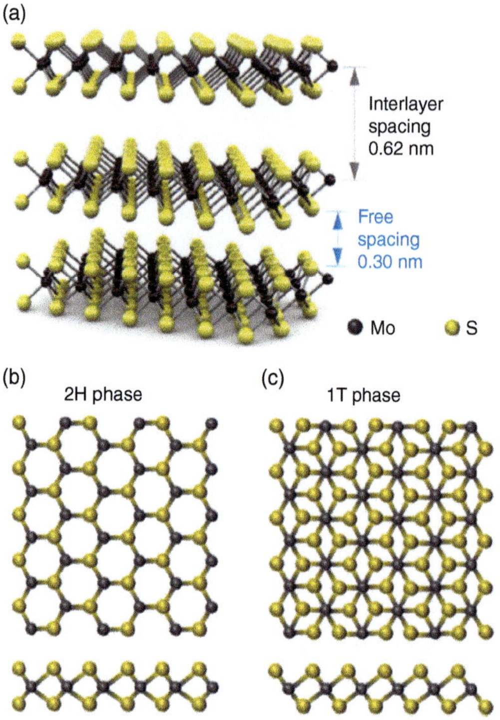

Figure 6.1 MoS_2 structure: (a) 3D illustration, (b) atomic positions in the 2H phase with triangular prism coordination, and (c) atomic positions in the 1T phase with octahedral coordination. Source: Wang and Mi [8]. Reproduced with permission of the American Chemical Society.

adjacent MoS_2 layers. Among bulk MoS_2 materials, natural molybdenite is typically a semiconductor with a thermodynamically stable 2H phase, whereas the 1T polymorph exhibits a metallic nature and does not exist under normal atmosphere [10]. Moreover, single- or few-layer MoS_2 nanosheets (MoS_2-NS) can be obtained using general physiochemical exfoliation methods because of the weak van der Waals forces between the layers. MoS_2 is of significant interest because of its physical, chemical, and mechanical properties. MoS_2 has a good electrical conductivity of ~1000 S cm^{-1} [11] and a carrier mobility of at least 200 cm^2 V^{-1} s^{-1}, which is similar to that of graphene ribbons [12]. Moreover, bulk MoS_2 has an indirect bandgap of 1.23 eV, whereas monolayers exhibit a direct bandgap of 1.8 eV [13, 14]. In addition to its excellent electronic properties, MoS_2 exhibits superior mechanical properties. The bending modulus of MoS_2 is higher than that of graphene by a factor of 7, which provides resistance against buckling. In addition, MoS_2 has a high Young's modulus of ~0.2 TPa and a fracture stress of more than 10 GPa [15]. Owing to these properties, MoS_2 has been extensively studied in various application areas, including supercapacitor electrodes, high-performance sensors, energy conversion, light-emitting diodes, and hydrogen evolution [16]. In addition, MoS_2 has shown excellent EMI shielding performance, which is attributed to its high electrical conductivity, good dielectric losses, and large specific surface area that attenuate incident EMWs by increasing the propagation pathways inside the sample.

Prasad et al. fabricated a $CoFe_2O_4$ nanoparticle-decorated MoS_2–reduced graphene oxide (rGO) nanocomposite to achieve improved shielding performance

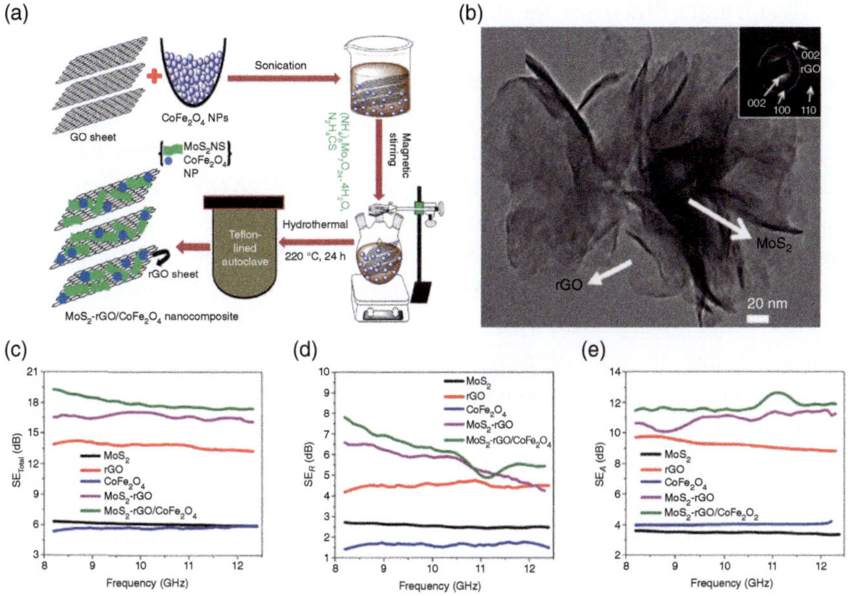

Figure 6.2 (a) Schematic of the synthesis of MoS_2–rGO and the MoS_2–rGO/$CoFe_2O_4$ nanocomposite. (b) TEM images of MoS_2–rGO. Frequency dependence of the EMI shielding performance: (c) SE_T, (d) SE_R, and (e) SE_A of MoS_2, rGO, $CoFe_2O_4$, MoS_2–rGO, and the MoS_2–rGO/$CoFe_2O_4$ nanocomposite in the X-band range. Source: Prasad et al. [17]. Reproduced with permission of the Royal Society of Chemistry.

[17]. The MoS_2–rGO/$CoFe_2O_4$ nanocomposite was synthesized by a two-step hydrothermal method, as shown in Figure 6.2a.

The MoS_2–rGO composite contained uniform three-dimensional (3D) flower-like microspheres of MoS_2 with a diameter of 1–2 μm, which were distributed randomly on the rGO sheets. Furthermore, the morphology was confirmed by transmission electron microscopy (TEM). In the TEM image (Figure 6.2b), the wrinkled rGO sheets are light gray and MoS_2 appears as a dark contrast. As shown in Figure 6.2c, the MoS_2-NS exhibited a total shielding effectiveness (SE_T) value of 6.33 dB, and the $CoFe_2O_4$ nanoparticles showed a similar low SE_T (5.37 dB) owing to their low conductivity. The rGO nanosheets with good electrical conductivity had a higher SE_T value of ~13.87 dB. Moreover, the MoS_2–rGO/$CoFe_2O_4$ nanocomposite showed a maximum SE_T of ~19.26 dB at a thickness of 1.4 mm, which was higher than the SE_T value of ~16.52 dB obtained for MoS_2–rGO. In addition, it was found that the SE_T of the MoS_2–rGO/$CoFe_2O_4$ nanocomposite decreased continuously with frequency, indicating that the $CoFe_2O_4$ nanoparticles in the composite provided superior shielding performance in the low-frequency range. As shown in Figure 6.2d, the shielding due to reflection (SE_R) values of MoS_2, rGO, $CoFe_2$O4, MoS_2–rGO, and MoS_2–rGO/$CoFe_2O_4$ were 2.71, 4.17, 1.52, 6.57, and 7.80 dB, respectively. Figure 6.2e represents the shielding due to absorption (SE_A) for the same samples, with values of 3.61, 9.69, 3.97, 10.60, and 12.62 dB, respectively. As the reflection contribution was much lower than the absorption contribution, absorption (SE_A) was the dominant contributor to the SE_T. Thus, the shielding performance of the MoS_2–rGO/$CoFe_2O_4$ composite can be attributed to moderate impedance matching, multiple reflections, and dipole polarization, which enhanced the dielectric and magnetic losses of the shield.

Unlike highly conductive nanomaterials such as graphene, carbon nanotubes (CNTs), and MXenes, semiconducting MoS_2 and its derivatives enhance impedance matching and improve the absorption of incident EMWs within the shield. Furthermore, several mechanisms contribute to the dissipation of EMWs, such as conduction and polarization losses. At higher frequencies, the EMI shielding performance is generally degraded, which can be compensated through the introduction of magnetic loss. For EMI absorption, magnetic losses such as natural resonance are prominent below 10 GHz, whereas eddy current losses and exchange resonance are dominant above 10 GHz. Menon et al. fabricated mussel-inspired self-healing polyurethane (PU) with magnetic Fe_3O_4 and MoS_2 as an efficient microwave absorber [18]. Schematic illustration of the synthesis of Fe_3O_4@MoS_2 has been shown in Figure 6.3a. Figure 6.3b,c show the appearance of Fe_3O_4@MoS_2, and Figure 6.3d shows the shape memory properties of the self-healing PU composites.

As shown in Figure 6.3e, pristine Fe_3O_4 and MoS_2 showed EMI SE values of −6.3 and −5.5 dB, respectively, at 18 GHz, whereas the hybrid Fe_3O_4@MoS_2 nanoparticles had an EMI SE value of −11.6 dB, i.e. a shielding efficiency of more than 90%, at a thickness of 5 mm. The multiwalled carbon nanotubes (MWNTs) showed a relatively high SE_T value of −23.6 dB, whereas the MWNT/Fe_3O_4@MoS_2 composite exhibited an improved SE_T of −36.6 dB. The introduction of Fe_3O_4 and MoS_2 yielded absorption dominant contributions of 68% and 56%, respectively (Figure 6.3f,g). The absorption contribution achieved with Fe_3O_4@MoS_2 was ~62%, which was slightly lower than that of Fe_3O_4 but higher than that of MoS_2. However, the MWNTs exhibited an absorption contribution of 87%, despite their conducting nature. The synergistic combination of MWNTs and Fe_3O_4@MoS_2 yielded a superior absorption contribution of 96%. The properties of the MWNT/Fe_3O_4@MoS_2 composite, which showed a low skin depth of 1.2 mm (as shown in Figure 6.3h), were outstanding. The hybrid polymer and its nanocomposite also exhibited good mechanical properties with recoveries of 75% and 56% of the original tensile strength, respectively. In the general EMI shielding mechanism, incident EMWs are attenuated as they pierce the shield material, and the EMW energy is dissipated in the form of heat through various loss mechanisms. The above results validated the hypothesis that the power of the incident EMWs could be attenuated using the complex heterogeneous shield.

MoS_2 has also been used as a microwave absorber owing to its moderate electrical conductivity, dielectric loss, and polarization loss. To design an efficient microwave absorber, two requirements must be satisfied: (i) low impedance difference, with impedance matching between the shield and air, and (ii) EMW attenuation, with high EMW absorption and the dissipation of EMW energy within the shield. Thus, the combination of moderate electrical conductivity with magnetic and/or dielectric fillers has been suggested ideal for microwave absorption.

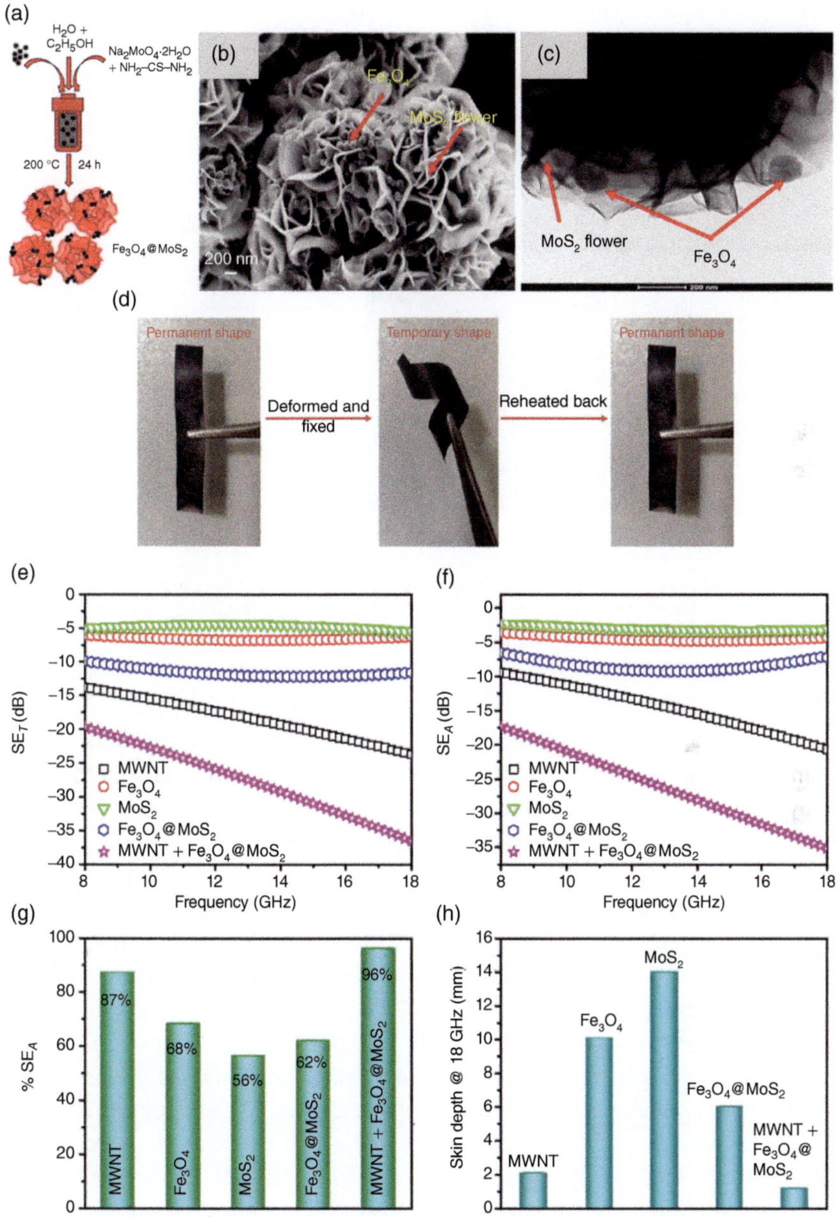

Figure 6.3 (a) Schematic of the synthesis of Fe_3O_4@MoS_2. (b) Scanning electron microscopy (SEM) image and (c) TEM image of Fe_3O_4@MoS_2. (d) Shape memory properties of the composite. (e) SE_T, (f) SE_A, and (g) % SE due to absorption and reflection and (h) skin depth. Source: Menon et al. [18]. Reproduced with permission of the American Chemical Society.

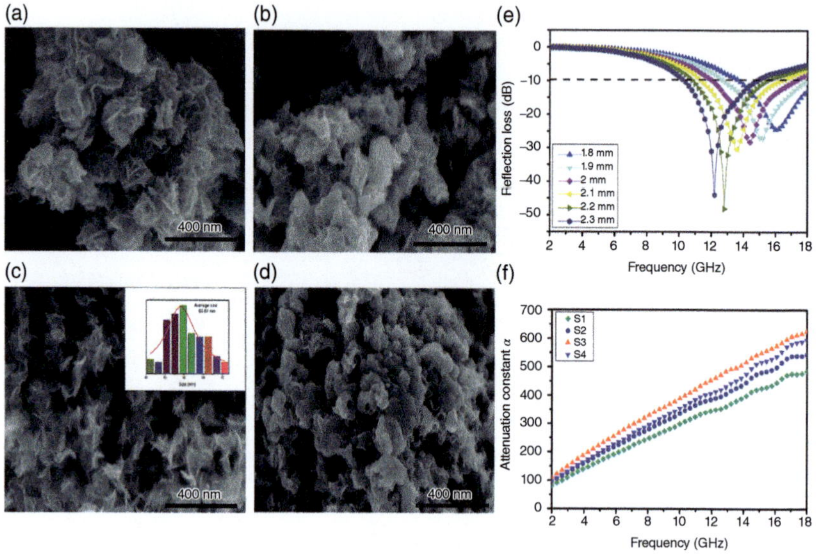

Figure 6.4 SEM images of MoS_2 nanosheets prepared at different hydrothermal temperatures: (a) S1 (160 °C), (b) S2 (170 °C), (c) S3 (180 °C), and (d) S4 (190 °C). (e) RL curves of MoS_2 (prepared at a hydrothermal temperature of 180 °C) at thicknesses of 1.8–2.3 mm in the frequency range of 2–18 GHz. (f) Attenuation constant (α) of the MoS_2 samples. Source: Liang et al. [19]. Reproduced with permission of the Royal Society of Chemistry.

Liang et al. synthesized 2D MoS_2-NS through a hydrothermal method for microwave absorption applications [19]. Four different compositions were synthesized at four different temperatures. Figure 6.4a–d shows the SEM images of MoS_2 synthesized at 160 °C, 170 °C, 180 °C, and 190 °C, respectively. Different morphologies were observed at different temperatures, with randomly distributed 2D sheet-like structures obtained at 180 °C. The MoS_2-NS showed a reflection loss (RL) value of −47.8 dB at a thickness of 2.2 mm with an effective absorption bandwidth (attenuation < −10 dB) of 4.5 GHz (11–15.5 GHz; see Figure 6.4e). Moreover, an effective bandwidth of 5.2 GHz was achieved at a thickness of 1.9 mm. Thus, the MoS_2-NS exhibited outstanding microwave absorption performance with high efficiency and broad bandwidths at low thicknesses. The attenuation constant (α) represents the integral attenuation ability, and a high α yields high dielectric and magnetic losses. As shown in Figure 6.4f, MoS_2 sample S3 (180 °C) exhibited high α values throughout the investigated frequency range. Thus, the excellent microwave absorption performance was attributed to high electrical conductivity and dielectric polarization losses.

Ning et al. fabricated few-layered MoS_2-NS from bulk MoS_2 via a top-down exfoliation method [20]. MoS_2-NS was then annealed at 150 °C for three hours, as shown in Figure 6.5a. Various MoS_2-NS and annealed MoS_2-NS composites were prepared using different concentrations of MoS_2 (30, 40, 50, and 60 wt%) in paraffin wax. An investigation of the dielectric properties revealed that the real permittivity of bulk MoS_2 decreased with increasing frequency but increased with increasing

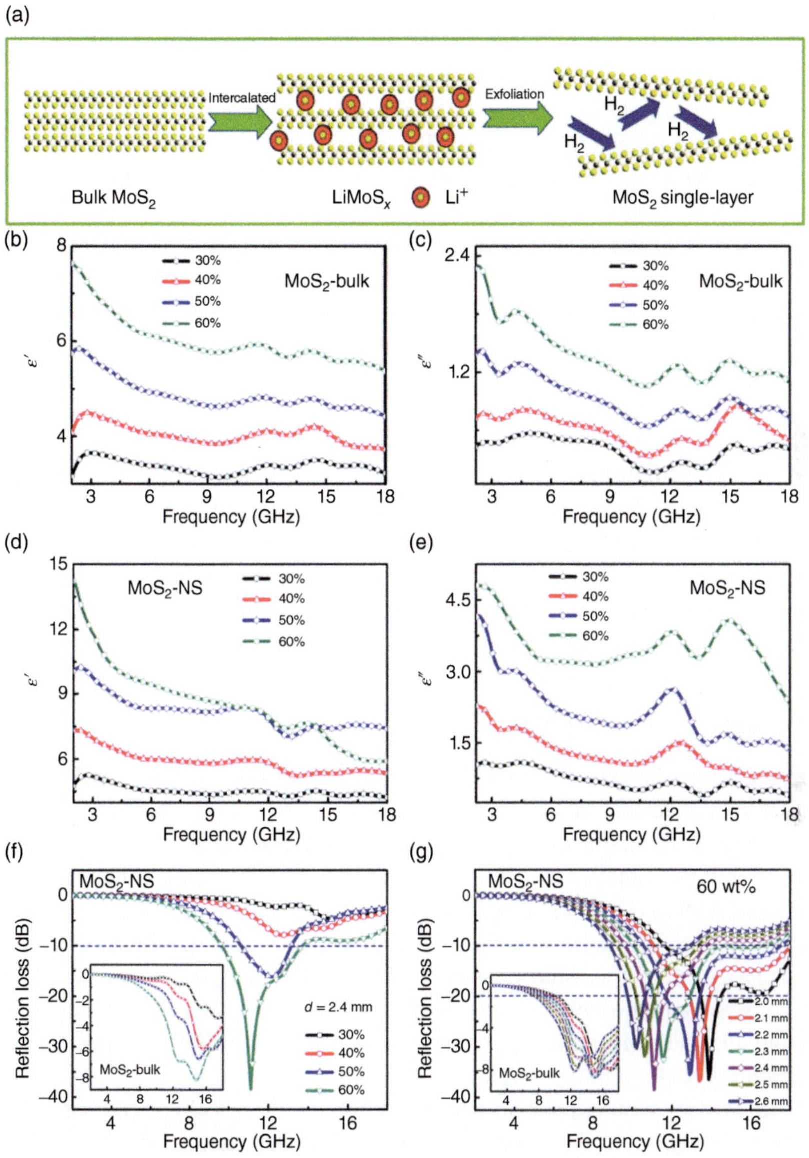

Figure 6.5 (a) Schematic of the exfoliation method for producing MoS_2-NS. ε' of (b) MoS_2-bulk/wax and (d) MoS_2-NS/wax with different loadings; ε'' of (c) MoS_2-bulk/wax and (e) MoS_2-NS/wax with different loadings. (f) RL of MoS_2-bulk/wax and MoS_2-NS/wax with different loadings at a thickness of 2.4 mm. (g) RL of MoS_2-bulk/wax and MoS_2-NS/wax with a loading of 60 wt% at different thicknesses. Source: Ning et al. [20]. Reproduced with permission of the Royal Society of Chemistry.

filler concentration (Figure 6.5b). In contrast, the imaginary permittivity decreased with increasing frequency (Figure 6.5c). Two relaxation peaks were observed at frequencies of 12.2 and 15 GHz, mainly arising from defect polarization. MoS_2-NS showed similar trends for ε' and ε'' to the MoS_2-bulk sample, but the values were almost two times higher (Figure 6.5d,e). The MoS_2-bulk sample possessed fewer defects, whereas the annealing process produced defects in MoS_2-NS, which can act as dipoles under an alternating electric field. Higher loadings (50 and 60 wt%) led to the formation of a conduction path, which facilitated the migration of electrons through the defects and interfaces in MoS_2-NS. The MoS_2-NS/wax composite with 60 wt% loading exhibited an RL value of −38.42 dB at a thickness of 2.4 mm, which was four times higher than that of the MoS_2-bulk/wax composite (−8.73 dB) (Figure 6.5f). Moreover, the MoS_2-NS/wax had an effective bandwidth of 4.1 GHz (9.6–13.76 GHz; see Figure 6.5g). The annealed MoS_2-NS/wax composite showed a maximum RL of −22.3 dB, which was almost three times higher than that of the MoS_2-bulk/wax composite at a comparable thickness. The obtained RL was attributed to the strong dielectric relaxation loss caused by defects, which enhanced the imaginary permittivity (Figure 6.5e). Furthermore, the formation of a conduction path dissipated the incident EMWs by ohmic or conduction loss. Finally, the high specific area of MoS_2-NS increased the propagation path for the EMWs within the sample owing to strong multiple reflections.

6.2.2 Tungsten Disulfide (WS_2)

WS_2, which also belongs to the TMDC family, has a layered hexagonal crystal structure similar to that of MoS_2, in which W atoms replace the Mo atoms in the trigonal coordination sphere. Bulk WS_2 has a dark gray color. Bulk WS_2 has an indirect electronic bandgap of 1.35 eV [21], whereas WS_2 monolayers have a direct bandgap of 2.05 eV [21, 22]. WS_2 has significant dielectric properties owing to abundant defects, dislocations, and a large surface area [23–25], with an electrical conductivity of 6.7 $S\,cm^{-1}$ and a carrier mobility of 0.2 $cm^2\,V^{-1}\,s^{-1}$ [12, 26]. In addition, WS_2 has excellent chemical stability and modifiable electronic properties from metallic to semiconducting, which provides great prospects for the field of EMI shielding [27].

Pristine WS_2 exhibits a poor SE because of its low electrical conductivity [28]. Therefore, Zhang et al. designed a self-assembled architecture of WS_2–rGO to improve the EMI shielding performance [29]. A new growth method was employed to obtain nanosheets, in which the guest phase was converted into the host phase. The growth process and mechanism of the WS_2–rGO nanosheets are depicted in Figure 6.6a. Figure 6.6b–d shows the TEM images of the samples at different growth stages. Wrinkles in rGO (dark brown areas in the TEM images marked with cyan dotted lines) were introduced as the reaction progressed. A 1.5 mm thick wax composite containing 60 wt% of the synthesized material was prepared and the shielding properties were analyzed. The EMI shielding performance of the WS_2–rGO composite depended on the rGO concentration and the frequency as shown in Figure 6.6e–h. As the frequency or rGO concentration increased, SE_T first increased and then decreased. The best EMI SE value of 32 dB was achieved

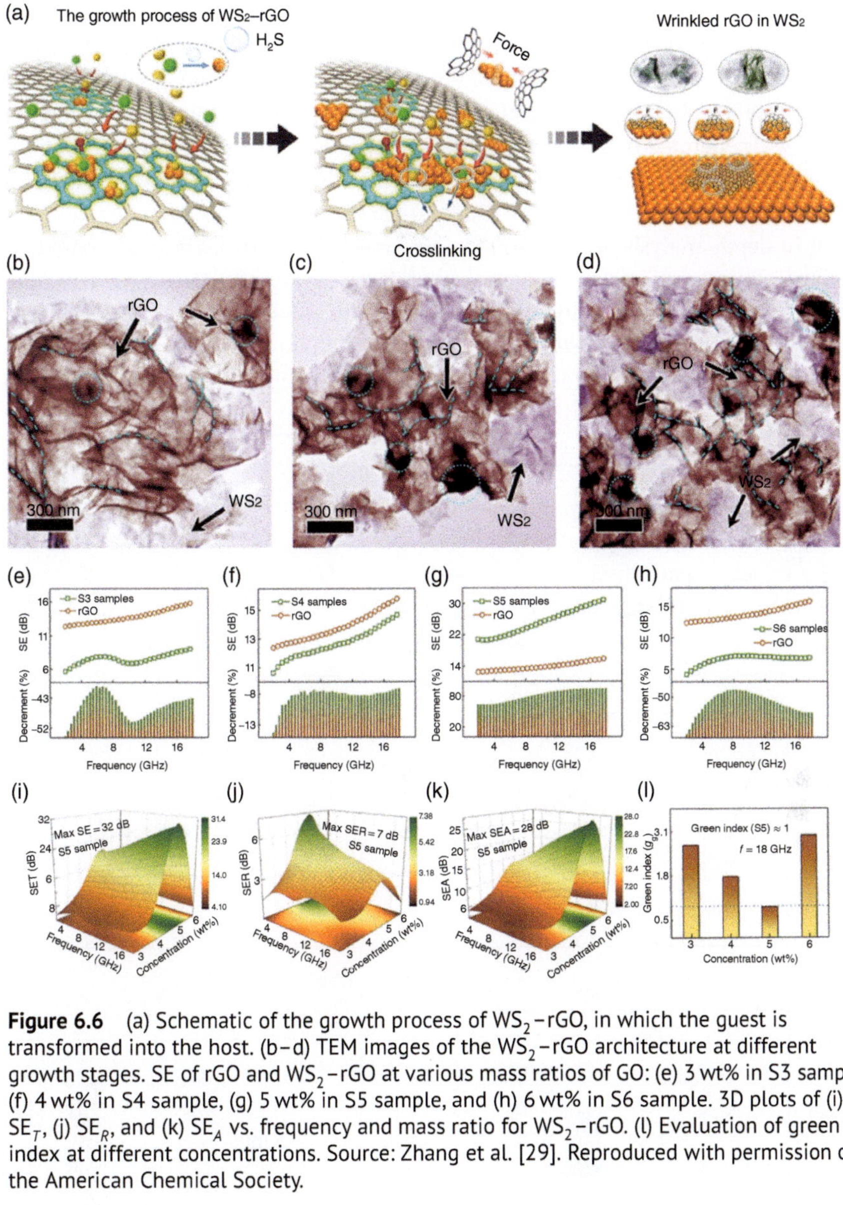

Figure 6.6 (a) Schematic of the growth process of WS_2–rGO, in which the guest is transformed into the host. (b–d) TEM images of the WS_2–rGO architecture at different growth stages. SE of rGO and WS_2–rGO at various mass ratios of GO: (e) 3 wt% in S3 sample, (f) 4 wt% in S4 sample, (g) 5 wt% in S5 sample, and (h) 6 wt% in S6 sample. 3D plots of (i) SE_T, (j) SE_R, and (k) SE_A vs. frequency and mass ratio for WS_2–rGO. (l) Evaluation of green index at different concentrations. Source: Zhang et al. [29]. Reproduced with permission of the American Chemical Society.

for the S5 sample with an rGO loading of only 5 wt%. 3D plots of EMI SE_T vs. frequency and the concentration of rGO have been shown in Figure 6.6i. The variation in EMI SE was attributed to the electrical conductivity and relaxation losses. The variations in SE_R and SE_A with frequency and mass ratio are shown in Figure 6.6j,k, respectively. SE_A was directly correlated with the frequency, whereas SE_R showed an inverserelation. As the frequency increased, the dielectric constant of the composite decreased, resulting in a decrease in reflectivity. The EMI shielding performance was attributed to high dielectric losses and multiple reflections. Furthermore, interfacial polarization in an alternating electric field owing to the difference in the surface conductivities of WS_2 and rGO, which can be considered as a capacitive structure, dissipated the absorbed EMWs in the form of heat. The low or moderate electrical conductivity of WS_2 aided impedance matching. The defects and dislocations in the structure also contributed to the dielectric losses within the shield and generated polarization and relaxation losses. Moreover, the morphology increased the barrier for the propagation path of the absorbed EMWs. Figure 6.6l shows evaluation of green index at different rGO concentrations.

An in-depth analysis of the EMI shielding mechanism was carried out based on various electromagnetic parameters, as depicted in Figure 6.7a–d. As discussed above, the WS_2–rGO architecture facilitated the dissipation of EMW energy owing to high dielectric losses and scattering, and these losses can be divided into conduction and relaxation losses. In the electromagnetic field, directional carriers

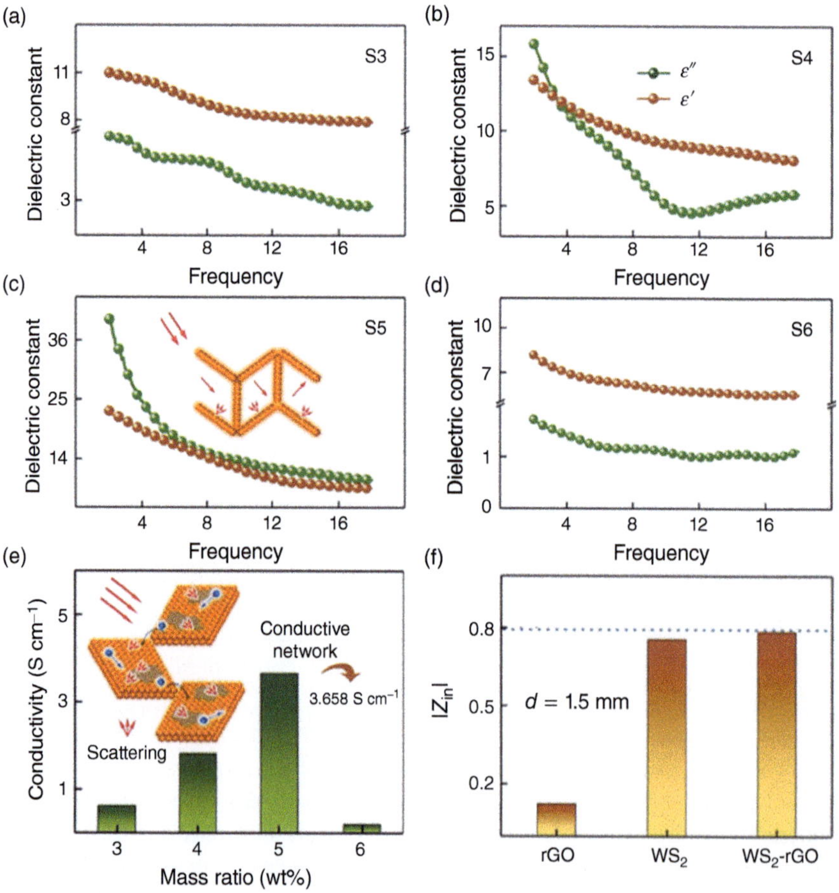

Figure 6.7 Real and imaginary permittivity of (a) S3 (GO/WS_2 mass ratio of 3), (b) S4 (GO/WS_2 mass ratio of 4), (c) S5 (GO/WS_2 mass ratio of 5), and (d) S6 (GO/WS_2 mass ratio of 6). (e) Conductivity of WS_2–rGO architecture incorporated in wax. (f) Z_{in} of rGO, WS_2, and WS_2–rGO at a thickness of 1.5 mm. Source: Zhang et al. [29]. Reproduced with permission of the American Chemical Society.

formed an electric current and converted this electrical energy into thermal energy. Moreover, the EMW energy was dissipated effectively at high frequencies owing to the slow dipole motion. In addition, interfacial polarization occurred in an alternating electric field owing to the difference in the surface conductivities of WS_2 and rGO, which can be considered as a capacitive structure and converted the absorbed EMW energy into thermal energy. The complex permittivity decreased with increasing frequency, whereas it initially increased with the concentration of rGO before decreasing at higher loadings, as shown for the S6 sample in Figure 6.7d. As depicted in Figure 6.7e, the conductive network formed between the WS_2–rGO nanosheets promoted electron transfer by a hopping mechanism to further enhance the EMW attenuation. However, in the case of S6, the reduction of graphene oxide (GO) was incomplete, which affected the formation of WS_2. In this case, the conduction and relaxation losses were reduced because the multilayered structure and interfaces were poorly formed. Furthermore, impedance matching was also an important factor [30]. As shown in Figure 6.7f, the impedance matching of the WS_2–rGO architecture was enhanced compared to that of pristine rGO and WS_2, which can be ascribed to the higher amount of terminal groups, defects, and interfaces in the WS_2–rGO architecture. Moreover, relaxation sites significantly improved the absorption of incident EMWs by suppressing secondary reflections and dissipating the electromagnetic energy into thermal energy.

6.2.3 Tantalum Disulfide (TaS_2)

TaS_2 is also a member of the TMDC family and is structurally similar to MoS_2 (Figure 6.8a–c). In the bulk state, TaS_2 has a 2H or 1T structure with two or one S–Ta–S sheets per unit cell, respectively. 2H-TaS_2 exhibits metallic behavior with superconductivity ($T_c = 0.8$ K). At room temperature, single-crystal 2H-TaS_2 has an electrical conductivity of 6.8×10^4 S cm^{-1}. TaS_2 with a layered structure, in which a layer of Ta atoms lies between two layers of sulfur atoms, exhibits a narrow energy gap of 0.3 eV [33]. Single-crystal 2H-TaS_2 exhibits a carrier mobility (μ) of 2.38 cm^2 V^{-1} s^{-1}. At a thickness of a few layers, TaS_2 exhibits enhanced superconductivity and gate-tunable phase transitions, which is a key prerequisite for certain applications such as electric oscillators, high-speed memory, and hydrogen generation catalysts [32]. The high electrical conductivity, good chemical stability in air, and excellent mechanical properties of TaS_2 make it an ideal choice for EMI shielding applications. The 2D layered structure provides improved interfaces or surfaces for multiple reflections and internal scattering. The structural, mechanical, and electrical properties of TaS_2 provide an alternative for designing absorption-dominant EMI shields.

Zong et al. fabricated flexible TaS_2/organic superlattice foils with excellent electrical properties, as shown in Figure 6.8d–h [32]. The TaS_2[hexylamine]$_x$[*N*-methylformamide]$_y$ ($TaS_2HA_xNMF_y$) foil was synthesized from a polycrystalline powder through an efficient and economical route involving ultrasonication-assisted exfoliation, mechanical grinding, and deposition by self-assembly. The $TaS_2HA_{0.371}$

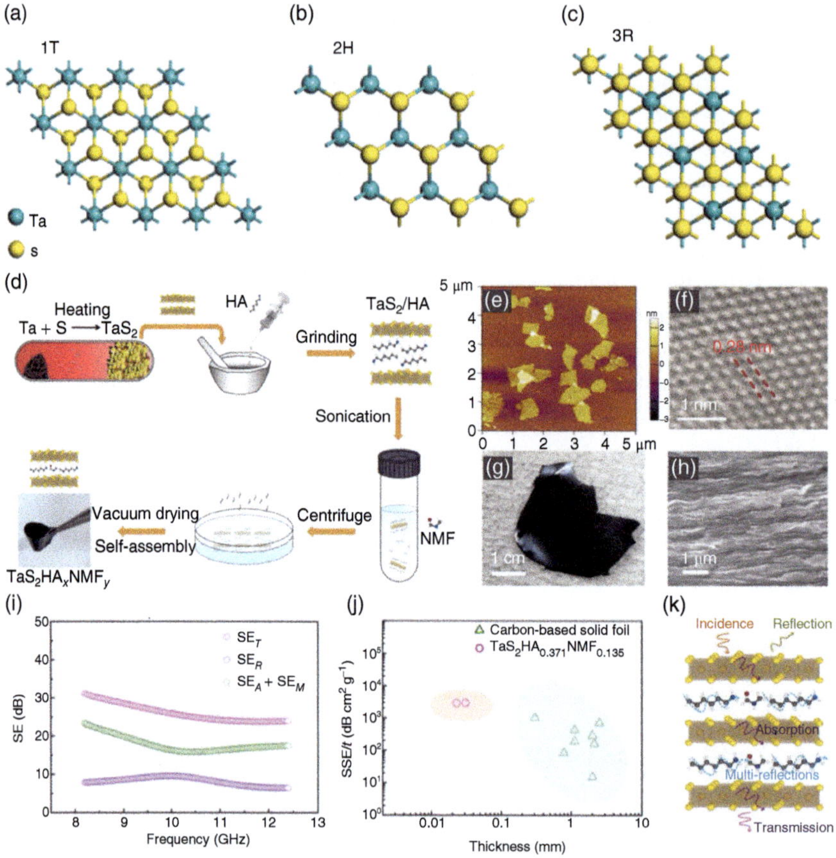

Figure 6.8 Schematics of different TaS_2 phases (top view): (a) 1T phase, (b) 2H phase (hexagonal symmetry), and (c) 3R phase (rhombohedral symmetry). Source: Feng et al. [31]. Reproduced with permission of the American Chemical Society. (d) Schematic of the solution-based synthesis processes of the hybrid $TaS_2HA_xNMF_y$ foil. (e) Atomic force microscopy (AFM) image of exfoliated TaS_2 nanosheets. (f) High-resolution TEM image of hybrid TaS_2HA_x nanosheet. (g) Surface appearance and (h) cross-sectional SEM image of hybrid $TaS_2HA_xNMF_y$ foil. (i) Frequency dependence of SE_T, SE_R, and ($SE_A + SE_M$) for the $TaS_2HA_{0.371}NMF_{0.135}$ foil. (j) SSE/t as a function of thickness for the as-prepared foils and carbon-based solid foils. (k) Schematic of the proposed EMI shielding mechanism in the TaS_2/organic superlattice foil. Source: Zong et al. [32]. Reproduced with permission of Wiley-VCH.

$NMF_{0.135}$ film exhibited an EMI SE of 31 dB at a frequency of 8.4 GHz and a thickness of 22 μm, as shown in Figure 6.8i. The contribution of SE_A and SE_M to the total EMI shielding of the film was 60–75%, which indicated an absorption-dominant shielding mechanism. For lightweight materials, it is important to account for the density and thickness when evaluating the EMI SE performance, and the absolute shielding effectiveness (SSE/t = EMI SE/density/thickness) is used for this purpose. As shown in Figure 6.8j, the SSE/t values of the mentioned film outperform those of carbon-based solid foils. The mechanical stability of the EMI shield is also an important factor. The synthesized film exhibited a bending modulus of 3.9 GPa at room temperature, which is analogous to those of an epoxy resin (3.3 GPa) and a graphene/epoxy resin composite (3.7 GPa) [34]. Figure 6.8k depicts the proposed EMI shielding mechanism for the film, in which the atomic sheets are consecutively stacked and separated by organic species. The absorbed EMWs underwent multiple reflections at the adjacent interfaces within the layers. Furthermore, dielectric losses were enhanced because the organic species acted as microcapacitors within the TaS_2 layers.

6.2.4 Hexagonal Boron Nitride (h-BN)

In the past few years, 2D h-BN has received significant attention because of its ultraflat surface, high heat dissipation, and highly stable structure. Furthermore, as h-BN is an electrical insulator, it can be used to alter the carrier mobility in various 2D materials such as graphene, MoS_2, and BP. 2D h-BN nanosheets are sp^2-hybridized and have a similar structure to graphene, consisting of an equal number of nitrogen and boron atoms alternatively arranged in a honeycomb structure. h-BN has an electronic bandgap in the range of 4.9–6.4 eV and a carrier mobility of 2300 $cm^2\ V^{-1}\ s^{-1}$ [35]. However, doping with certain elements minimizes the wide bandgap, which can result in a change from insulating to semiconducting properties. Modern electronics are highly integrated and miniaturized, which leads to equipment overheating and reduces reliability. Many studies have focused on attenuating EMWs through conduction and ohmic losses, where electromagnetic energy is dissipated in the form of heat. However, conventional conductive materials such as graphene and CNTs fail to provide sufficient thermal insulation and heat energy dissipation. To solve this problem, h-BN can be employed because of its high thermal conductivity (~280 $W\ m^{-1}\ K^{-1}$) and excellent electrical insulation performance [36, 37]. Moreover, the layered 2D morphology, semiconducting properties, excellent thermal conductivity, and ease of functionalization of h-BN can be exploited to design EMI shields with good electrical insulation and heat energy dissipation. However, hybrid composite materials should be considered to compensate for the insulating nature of h-BN.

In this direction, Zhang et al. used a layer-by-layer (LBL) approach to prepare a layered structure with excellent EMW shielding, good electrical insulation, and high thermal conductivity [38]. An LBL sandwich structure was fabricated by casting alternating GO/polymer and h-BN/polymer layers (Figure 6.9a–f). The hybrid structure possessed a core GO/polymer layer with h-BN/polymer as the skin layer. The fabricated architecture provided the dual advantage of establishing a thermal and conductive path for shielding in the in-plane direction and efficiently blocking the electrical conduction path in the vertical plane. The materials used in this film were maleic anhydride-grafted-styrene-ethylene/butylene-styrene block copolymer (SEBS-g-MAH), phosphorus-containing liquid crystalline copolyester (PHDDT), GO, and h-BN.

The EMI SE of three-layer films with different loadings of f-GO and a thickness of 233 μm in the X-band frequency range is shown in Figure 6.9g. The EMI SE value increased with increasing f-GO content, with a maximum EMI SE value

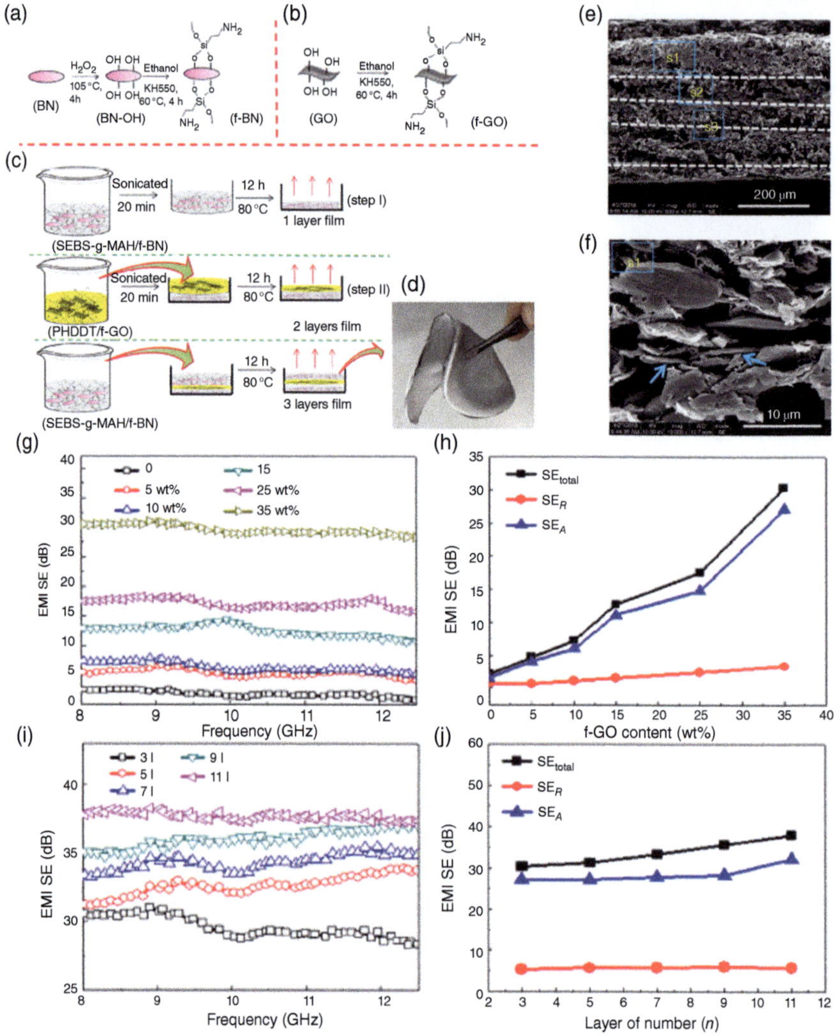

Figure 6.9 Schematic of the fabrication of the ordered multilayer film: (a) surface functionalization of h-BN particles (f-BN), (b) surface functionalization of GO particles (f-GO), and (c) preparation of the ordered multilayer film. (d) Optical photograph of a three-layer film. (e) SEM image of a five-layer film. (f) SEM image of the SEBS-g-MAH/f-BN layer (region s1). (g) EMI SE as a function of frequency for three-layer films with different f-GO contents. (h) SE_T, SE_A, and SE_R of three-layer films with different f-GO contents at a frequency of 8.2 GHz. (i) EMI SE as a function of frequency for ordered multilayer films with different numbers of layers. (j) SE_T, SE_A, and SE_R of ordered multilayer films with different numbers of layers at 8.2 GHz. Source: Zhang et al. [38]. Reproduced with permission of Elsevier.

of 29.7 dB achieved with 35 wt% f-GO. SE_A and SE_T both increased as the f-GO content increased, whereas the SE_R contribution was insignificant for all f-GO contents (Figure 6.9h). Figure 6.9i shows the EMI SE of aligned multilayer films with different numbers of layers. The EMI SE improved as the number of layers increased, and the hybrid structure with 11 layers containing 35 wt% f-GO exhibited the maximum EMI SE value of 37.92 dB. As shown in Figure 6.9j, both SE_R and SE_A increased as the number of layers increased.

6.2.5 Black Phosphorus (BP)

BP is a new 2D material that has gained significant attention in both academic research and industrial applications owing to its high charge carrier mobility, controllable direct bandgap properties, and unique in-plane anisotropic structure. Phosphorus belongs to group V of the periodic table and has four major allotropes: white phosphorus (WP), red phosphorus (RP), violet phosphorus (VP), and BP [39]. The bandgap of BP is always direct and ranges from 0.3 to 2.0 eV [40, 41], which provides a new alternative between zero-gap graphene and the large-gap TMDCs. Moreover, unlike graphene and TMDCs, BP exhibits a direct bandgap in its monolayer, few-layer, and bulk forms. BP possesses a puckered hexagonal structure with two different floor plane configurations: armchair (AC) and zigzag (ZZ) (Figure 6.10a,b) [42]. BP exhibits various unique properties such as structural anisotropy, strong mechanical anisotropy, and good thermal, electrical, and optical properties. For instance, BP has a carrier mobility of $1000\,cm^2\,V^{-1}\,s^{-1}$ and thermal conductivities of 110 and $35\,W\,m^{-1}\,K^{-1}$ (AC and ZZ forms, respectively) [44, 45].

Owing to issues related to the synthesis of BP, research on electromagnetic shielding and absorption remains in the early stages [46]. However, the large-scale synthesis of BP nanosheets by liquid-phase exfoliation (LPE) has recently been reported, thus facilitating the analysis of BP for EMI shielding applications and microwave absorption [47]. BP should be investigated as a microwave absorber to determine whether its semiconducting nature improves impedance matching and the attenuation of EMWs. Wu et al. reported that few-layer black phosphorus (FL-BP) synthesized by LPE exhibits high electromagnetic absorption in multiple frequency bands [43]. The RL curves of the composites were analyzed to determine the MWA performance. Composites with FL-BP at a filler loading ratio of 30 wt% achieved a wide bandwidth of up to 6.20 GHz at a thickness of 2.5 mm (Figure 6.10c). An RL_{min} value of −46.5 dB was observed in the Ku-band for the 2.2 mm thick sample (Figure 6.10d), whereas an RL_{min} value of −41.6 dB was observed in the X-band for the 2.7 mm thick sample (Figure 6.10e). As shown in Figure 6.10f,g, at 50 wt% loading of FL-BP, the composites also had an effective bandwidth in the S-band, with an RL_{min} value of −20.1 dB at a thickness of 6 mm (Figure 6.10h).

6.2.6 Copper Sulfide (CuS)

The Cu_xS $(1 < x < 2)$ system is well known for its excellent p-type conduction among narrow bandgap semiconductors, and there are five stable phases at room temperature: covellite (CuS), anilite ($Cu_{1.75}S$), digenite ($Cu_{1.8}S$), djurleite ($Cu_{1.95}S$), and chalcocite (Cu_2S). Copper monosulfide (CuS) is a stacked semiconductor crystal

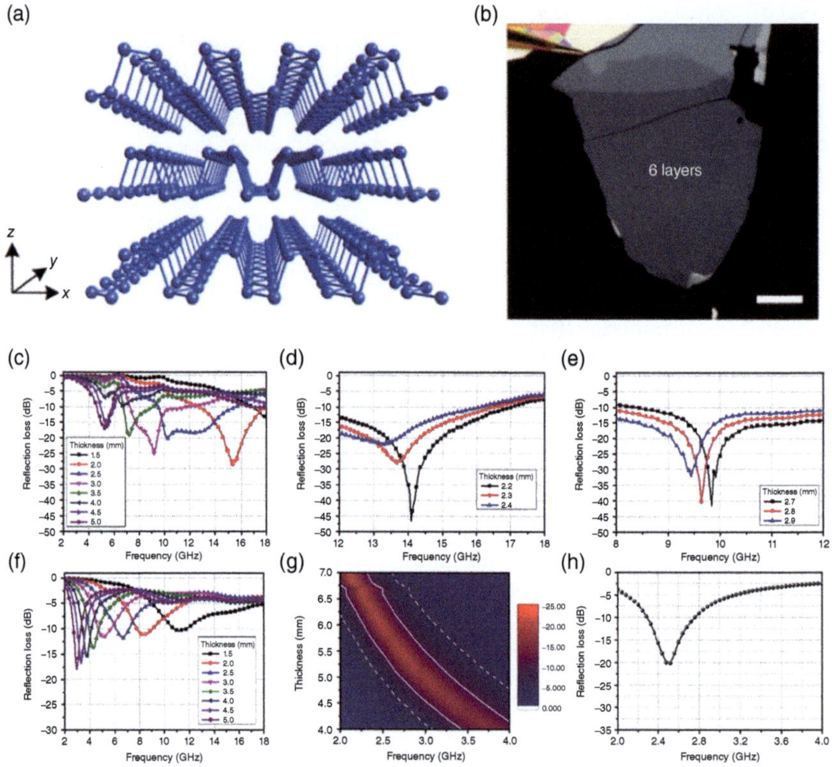

Figure 6.10 (a) Lattice structure of few-layer BP showing the puckered hexagonal crystal with two nonequivalent directions: AC (x-direction) and ZZ (y-direction). (b) Optical image of a representative six-layer sample. Scale bar: 20 mm. Source: Zhang et al. [42]. Reproduced with permission of Springer Nature. RL curves of a composite loaded with 30 wt% FL-BP: (c) thicknesses of 1.5–5 mm at frequencies of 2–18 GHz, (d) thicknesses of 2.2, 2.3, and 2.4 mm in the Ku-band and (e) thicknesses of 2.7, 2.8, and 2.9 mm in the X-band. RL curves of a composite loaded with 50 wt% FL-BP: (f) thicknesses of 1.5–5 mm at frequencies of 2–18 GHz, (g) 3D plot in the S-band for thicknesses of 4.0–7.0 mm (white dashed line: −5 dB; white line: −10 dB), and (h) thickness of 6 mm in the S-band. Source: Wu et al. [43]. Reproduced with permission of Elsevier.

with a direct bandgap of ~2 eV at room temperature [48]. CuS nanoplates have a layered crystal structure with a hexagonal lattice, as shown in Figure 6.11a. As the atomic layers are attached by weak van der Waals forces, thin and single layers can be obtained through exfoliation, as shown in an optical image in Figure 6.11b, making CuS a potential candidate for fabricating 2D materials and heterostructures [49]. CuS has attracted significant interest in the field of materials science owing to its excellent optical, electronic, and chemical properties, and the ease of synthesizing different CuS morphologies, such as hexagonal nanoplates, nanoflakes, and nanotubes, has broadened its applications in electronics, mechanical devices, and other fields [50].

Liu et al. investigated the microwave absorption performance of CuS nanoflakes aligned perpendicularly on magnetically decorated graphene [50]. Moreover, the

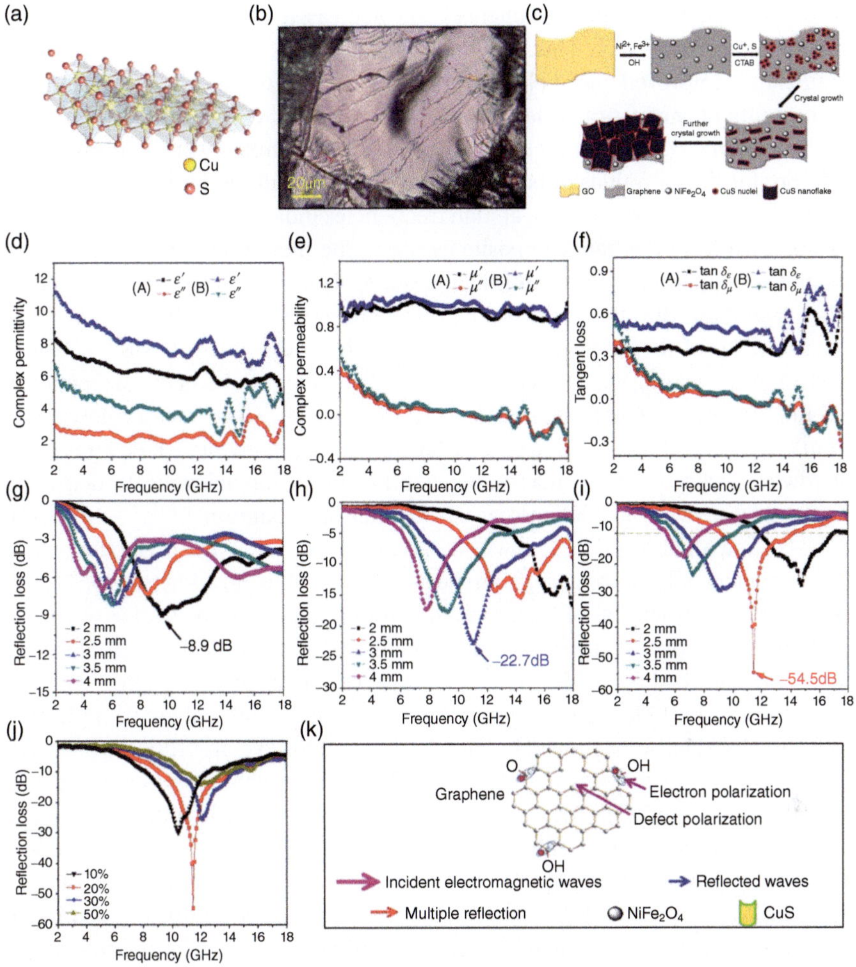

Figure 6.11 (a) Crystal structure of CuS. Gold and orange balls represent Cu and S atoms, respectively. (b) Optical microscopy image of a bulk CuS crystal. Scale bar: 20 μm. Source: Miao et al. [49]. Reproduced with permission of the American Chemical Society. (c) Schematic of the formation of magnetically decorated graphene@CuS. (d) Relative permittivity, (e) relative permeability, and (f) tangent loss of (A) magnetically decorated graphene and (B) magnetically decorated graphene@CuS with 0.5 mmol CTAB. RL curves of (g) graphene, (h) magnetically decorated graphene, and (i) magnetically decorated graphene@CuS with a 20 wt% loading. (j) RL curves of magnetically decorated graphene@CuS with different loadings at a thickness of 2.5 mm. (k) Possible microwave absorption mechanism in magnetically decorated graphene@CuS. Source: Liu et al. [50]. Reproduced with permission of Elsevier.

effect of cetyltrimethylammonium bromide (CTAB) on the morphology of the CuS nanoflakes was studied. Figure 6.11c shows the fabrication of the magnetically decorated graphene@CuS composite. As shown in Figure 6.11d, within the 2–18 GHz range, the ε' and ε'' values of magnetically decorated graphene decreased marginally in the ranges of 4.4–8.7 and 1.8–3.5, respectively, whereas those of the magnetically decorated graphene@CuS decreased in the ranges of 6.7–11.7 and 2.4–6.6, respectively. This behavior indicated that the introduction of CuS significantly improved the dielectric constant. Figure 6.11e shows the complex permeability of magnetically decorated graphene and the magnetically decorated graphene@CuS composite. The μ' values of both samples were equivalent. In addition, for the μ'' values of both samples, a broad multiple resonance peak was observed, and the negative value at high frequencies indicated that magnetic energy was released from the nanocomposite owing to the motion of the charge. For the calculated tangent loss (Figure 6.11f), the tan δ_ε value was higher than the tan δ_μ value, signifying that the magnetic loss was low. As shown in Figure 6.11g, the maximum RL of the graphene composite at a thickness of 2 mm was −8.9 dB at 9.5 GHz. After coating with $NiFe_2O_4$ particles, the maximum RL of the magnetically decorated graphene composite reached −22.7 dB at 11.1 GHz with a thickness of 3 mm (Figure 6.11h), indicating that better impedance matching yielded improved microwave absorption performance. Figure 6.11i shows that magnetically decorated graphene@CuS with a thickness of 2.5 mm had a maximum RL of −54.5 dB at 11.4 GHz and an absorption bandwidth of 4.5 GHz (9.7–14.2 GHz). The theoretical RL of magnetically decorated graphene@CuS with filler loadings of 10, 20, 30, and 50 wt% at a thickness of 2.5 mm is shown in Figure 6.11j. A loading of 20 wt% exhibited the strongest microwave absorption and is thus ideal for an absorber. Increasing the loading of the nanocomposite could result in a highly complex dielectric constant, leading to impedance mismatch and enhanced reflection from the absorber surface, which degrades the microwave absorption properties. Figure 6.11k shows the possible microwave absorption mechanism in magnetically decorated graphene@CuS. The introduction of magnetic particles into the graphene sheets and CuS nanoflakes enhanced interfacial polarization, dipole polarization, and ionic polarization. Moreover, the residual defects and terminal groups in graphene contributed to defect polarization relaxation, electron polarization, and relaxation processes that enhanced the microwave absorption properties of the composite.

6.2.7 Metal–Organic Frameworks (MOFs)

MOFs, also called porous coordination polymers (PCPs), synthesized by combining metal ions and organic ligands, are characterized by tunable structural design, the possibility of fine-tuning, and a uniform pore structure [51]. Magnetic metal nanoparticles embedded in porous carbon derived from MOFs with magnetic elements as coordination metals have been widely used as microwave absorbers because of their ease of synthesis, good electrical conductivity, and large specific surface areas. However, the literature suggests that MOF-based absorbers should have loadings higher than 40 wt% to realize good microwave absorption, as these particulate absorbents lack the ability to form conductive networks in the matrix [52]. Thus, strategies for designing conductive and/or magnetic MOFs as well as the conductive and magnetic properties of MOFs and their physical origins have been systematically studied [53, 54]. Multiphase materials, such as MOFs, with moderate electrical conductivity and high magnetic or dielectric losses are ideal for inhibiting electromagnetic pollution. MOFs can facilitate electron migration or hopping within the EMI shield, and heterogeneous interfaces can lead to polarization, thus enhancing EMW attenuation. Wang et al. reported a green method for synthesizing 2D Co/C nanosheets by directly carbonizing zeolitic imidazolate framework (ZIF-67) nanosheets using water instead of methanol as the solvent. The effects of the pyrolysis conditions on the microstructure, degree of graphitization, magnetic properties, and microwave absorption properties of the 2D Co/C composite materials were determined [52]. The prepared 2D Co/C composites were added to paraffin to evaluate the microwave absorption characteristics. Figure 6.12a–c shows the morphologies of the samples obtained by the pyrolysis of ZIF-67 at different temperatures (500, 700, and 900 °C). The T-Co/C-700 and T-Co/C-900 composites exhibited pore formation, which was attributed to the loss of the organic components from the ZIF-67 nanosheet precursor.

As shown in Figure 6.12d,e, for the wax composite with 10 wt% T-Co/C, the complex permittivity (ε' and ε'') in the range of 2–18 GHz increased with increasing pyrolysis temperature. Furthermore, the dielectric loss tangent ($\tan \delta_\varepsilon = \varepsilon''/\varepsilon'$) values of the T-Co/C-700 and T-Co/C-900 composites were higher than those of the T-Co/C-500 composites (Figure 6.12h), signifying a strong dielectric loss ability over the entire frequency range. Thus, the presence of defects and dipoles in the T-Co/C composite generated polarization to dissipate electromagnetic energy as thermal energy in the presence of alternating electromagnetic fields [20]. Moreover, the free charge could easily build up in the Co@C heterojunctions in the T-Co/C composite as well as at the multiple interfaces between the T-Co/C composite and the paraffin matrix, resulting in interfacial polarization, associated relaxation processes, and the attenuation of incident EMWs. In addition, the three samples only showed slightly different μ' and μ'' values, which means they had similar magnetic loss characteristics, as confirmed by the magnetic loss tangent (Figure 6.12i). Figure 6.12j–l shows the calculated RL values for T-Co/C-500, T-Co/C-700, and T-Co/C-900 at thicknesses of 2–4 mm. For T-Co/C-500 (Figure 6.12j), the maximum RL was −13.9 dB at 17.1 GHz with an effective bandwidth (RL < −10 dB) of 2.2 GHz (15.8–18.0 GHz) at a thickness of 2.5 mm. For T-Co/C-700 (Figure 6.12k), the maximum RL increased to −39.8 dB at a thickness of 2.2 mm, and accordingly, the effective bandwidth was extended to 5.3 GHz (11.0–16.3 GHz). For T-Co/C-900 (Figure 6.12l), the maximum RL value reached −47.8 dB at a thickness of 2.9 mm, but the effective bandwidth was reduced to 3.8 GHz (6.6–10.4 GHz). The excellent microwave absorption achieved with only 10 wt% loading of T-Co/C was attributed to the formation of an effective conductive path that facilitated electron migration and hopping. Moreover, the 2D morphology enhanced multiple reflections and increased the EMW propagation paths within the shield. The synergistic combination of these processes improved the attenuation of electromagnetic energy and dielectric loss.

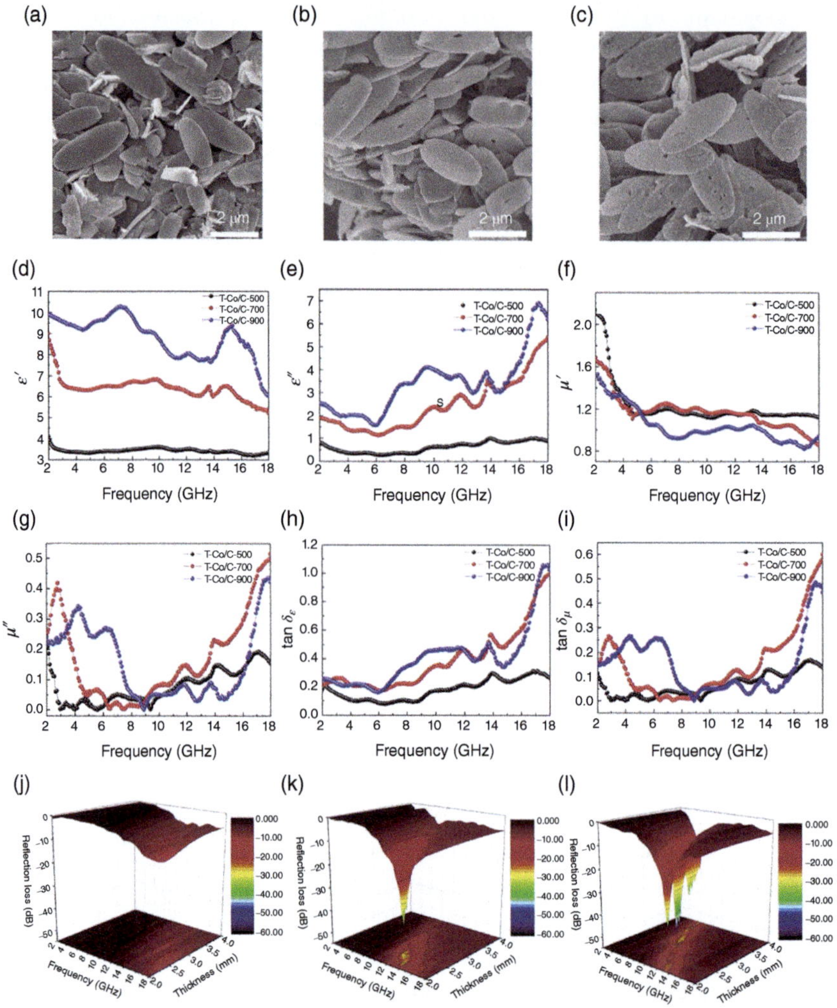

Figure 6.12 SEM images of (a) T-Co/C-500, (b) T-Co/C-700, and (c) T-Co/C-900. (d) Real part of permeability, (e) imaginary part of permeability, (f) real part of permittivity, (g) imaginary part of permittivity, (h) dielectric loss tangent, and (i) magnetic loss tangent of T-Co/C-500, T-Co/C-700, and T-Co/C-900. 3D plots of the calculated RL values for (j) T-Co/C-500, (k) T-Co/C-700, and (l) T-Co/C-900 paraffin wax composites with different thicknesses in the frequency range of 2–18 GHz. Source: Wang et al. [52]. Reproduced with permission of Elsevier.

6.3 Summary

The two principal requirements for superior microwave absorption materials are (i) impedance matching between air and the shield and (ii) electromagnetic attenuation via effective EMW absorption and energy dissipation by the shield. To satisfy the basic requirements for ideal EMI shields, materials should have moderate conductivity, high dielectric losses, and magnetic losses. The carbon-based family of EMI shielding and microwave absorption materials is at the forefront of EMW attenuation. Although the high electrical conductivity and dielectric losses of these materials are advantageous for EMW attenuation, the resultant impedance mismatching severely degrades the electromagnetic absorption properties. As described in this chapter, various alternative 2D materials, such as MoS_2, WS_2, TaS_2, and h-BN, also demonstrate promising results for EMI shielding and microwave absorption. The moderate electrical conductivity of these materials facilitates EMW penetration into the shield owing to proper impedance matching. Thus, the moderate electrical conductivity, high thermal conductivity, large surface area, and ease of processing contribute to the EMI shielding performance of these materials. Moreover, their composites and hybrids show absorption-based EMI shielding mechanisms with excellent heat dissipation properties. Modern smart electronic systems require special materials with unique physical structures, lightweights, and environmental stability for EMI shielding. Various approaches for absorbing electromagnetic radiation have been investigated, such as segregated structures, laminate films, and porous materials. However, 2D materials such as MoS_2, WS_2, TaS_2, and h-BN have not been explored to their limits. Further in-depth studies are required to assess their capabilities as EMI shielding materials, which may open a new chapter in EMI shielding strategies.

References

1 Chhowalla, M., Shin, H.S., Eda, G. et al. (2013). The chemistry of two-dimensional layered transition metal dichalcogenide nanosheets. *Nature Chemistry* 5 (4): 263–275.

2 Huang, X., Zeng, Z., and Zhang, H.J. (2013). Metal dichalcogenide nanosheets: preparation, properties and applications. *Chemical Society Reviews* 42 (5): 1934–1946.

3 Xu, M., Liang, T., Shi, M. et al. (2013). Graphene-like two-dimensional materials. *Chemical Reviews* 113 (5): 3766–3798.

4 Chhowalla, M., Liu, Z.F., and Zhang, H. (2015). Two-dimensional transition metal dichalcogenide (TMD) nanosheets. *Chemical Society Reviews* 44 (9): 2584–2586.

5 Lin, Y., Williams, T.V., and Connell, J.W. (2010). Soluble, exfoliated hexagonal boron nitride nanosheets. *Journal of Physical Chemistry Letters* 1 (1): 277–283.

6 Peng, Y., Li, Y.S., Ban, Y.J., and Jin, H. (2014). Metal-organic framework nanosheets as building blocks for molecular sieving membranes. *Science* 346 (6215): 1356–1359.

7 Rodenas, T., Luz, I., Prieto, G. et al. (2015). Metal–organic framework nanosheets in polymer composite materials for gas separation. 14 (1): 48–55.

8 Wang, Z. and Mi, B.J.E. (2017). Environmental applications of 2D molybdenum disulfide (MoS_2) nanosheets. *Environmental Science and Technology* 51 (15): 8229–8244.

9 Ai, K., Ruan, C., Shen, M., and Lu, L.J.A.F.M. (2016). MoS_2 nanosheets with widened interlayer spacing for high-efficiency removal of mercury in aquatic systems. *Advanced Functional Materials* 26 (30): 5542–5549.

10 Enyashin, A.N., Yadgarov, L., Houben, L., and Popov, I. (2011). New route for stabilization of 1T-WS_2 and MoS_2 phases. *The Journal of Physical Chemistry C* 115 (50): 24586–24591.

11 Pearce, C.I., Pattrick, R.A.D., and Vaughan, D.J. (2006). Electrical and magnetic properties of sulfides. *Reviews in Mineralogy and Geochemistry* 61: 127–180.

12 Velicky, M. and Toth, P.S. (2017). From two-dimensional materials to their heterostructures: an electrochemist's perspective. *Applied Materials Today* 8: 68–103.

13 Kobayashi, K. and Yamauchi, J. (1995). Electronic structure and scanning-tunneling-microscopy image of molybdenum dichalcogenide surfaces. *Physical Reviews B* 51 (23): 17085.

14 Yun, W.S., Han, S., Hong, S.C., and Kim, I.G. (2012). Thickness and strain effects on electronic structures of transition metal dichalcogenides: 2H-MX_2 semiconductors (M = Mo, W; X = S, Se, Te). *Physical Reviews B* 85 (3): 033305.

15 Ghobadi, N. (2017). A comparative study of the mechanical properties of multilayer MoS_2 and graphene/MoS_2 heterostructure: effects of temperature, number of layers and stacking order. *Current Applied Physics* 17 (11): 1483–1493.

16 Wang, X.X., Zhang, W.L., Ji, X.Q. et al. (2016). 2D MoS_2/graphene composites with excellent full Ku band microwave absorption. *RSC Advances* 6 (108): 106187–106193.

17 Prasad, J., Singh, A.K., Haldar, K.K. et al. (2019). $CoFe_2O_4$ nanoparticles decorated MoS_2-reduced graphene oxide nanocomposite for improved microwave absorption and shielding performance. *RSC Advances* 9 (38): 21881–21892.

18 Menon, A.V., Madras, G., and Bose, S. (2019). Mussel-inspired self-healing polyurethane with "flower-like" magnetic MoS_2 as efficient microwave absorbers. *ACS Applied Polymer Materials* 1 (9): 2417–2429.

19 Liang, X., Zhang, X.; Liu, W.; Tang, D.; Zhang, B.; Ji, Guangbin, A simple hydrothermal process to grow MoS_2 nanosheets with excellent dielectric loss and microwave absorption performance. *Journal of Materials Chemistry C* 2016. 4(28): p. 6816–6821.

20 Ning, M.-Q., Lu, M.-M., Li, J.-B., and Chen, Z. (2015). Two-dimensional nanosheets of MoS_2: a promising material with high dielectric properties and microwave absorption performance. *Nanoscale* 7 (38): 15734–15740.

21 Baglio, J.A., Calabrese, G.S., Kamieniecki, E., and Kershaw, R. (1982). Characterization of N-type semiconducting tungsten disulfide photo-anodes in aqueous and non-aqueous electrolyte-solutions – photooxidation of halides with high-efficiency. *Journal of the Electrochemical Society* 129 (7): 1461–1472.

22 Gutierrez, H.R., Perea-Lopez, N., Elias, A.L. et al. (2013). Extraordinary room-temperature photoluminescence in triangular WS_2 monolayers. *Nano Letters* 13 (8): 3447–3454.

23 Zhang, Y., Onga, M., Qin, F. et al. (2018). Optoelectronic response of a WS_2 tubular pn junction. *2D Materials* 5 (3): 035002.

24 Xia, D.D., Gong, F., Pei, X. et al. (2018). Molybdenum and tungsten disulfides-based nanocomposite films for energy storage and conversion: a review. *Chemical Engineering Journal* 348: 908–928.

25 Ding, W., Hu, L., Liu, Q. et al. (2018). Structure modulation induced enhancement of microwave absorption in WS_2 nanosheets. *Applied Physics Letters* 113 (24): 243102.

26 Balach, J., Linnemann, J., Jaumann, T., and Giebeler, L. (2018). Metal-based nanostructured materials for advanced lithium–sulfur batteries. *Journal of Materials Chemistry A* 6 (46): 23127–23168.

27 Sheng, Y., Chen, T., Lu, Y. et al. (2019). High-performance WS_2 monolayer light-emitting tunneling devices using 2D materials grown by chemical vapor deposition. *ACS Nano* 13 (4): 4530–4537.

28 Wang, Q.H., Kalantar-Zadeh, K., Kis, A., and Coleman, J.N. (2012). Electronics and optoelectronics of two-dimensional transition metal dichalcogenides. *Nature Nanotechnology* 7 (11): 699–712.

29 Zhang, D.-Q., Liu, T.-T., Shu, J.-C. et al. (2019). Self-assembly construction of WS_2–rGO architecture with green EMI shielding. *ACS Applied Materials & Interfaces* 11 (30): 26807–26816.

30 Wen, B., Cao, M., Lu, M. et al. (2014). Reduced graphene oxides: light-weight and high-efficiency electromagnetic interference shielding at elevated temperatures. *Advanced Materials* 26 (21): 3484–3489.

31 Feng, Y., Gong, S., Du, E. et al. (2018). 3R TaS_2 surpasses the corresponding 1T and 2H phases for the hydrogen evolution reaction. *The Journal of Physical Chemistry C* 122 (4): 2382–2390.

32 Zong, P.A., Yoo, D., Zhang, P. et al. (2020). Flexible foil of hybrid TaS_2/organic superlattice: fabrication and electrical properties. *Small* 16 (15): 1901901.

33 Sunilkumar, A., Manjunatha, S., Chethan, B., and Ravikiran, Y.T. (2019). Polypyrrole–tantalum disulfide composite: an efficient material for fabrication of room temperature operable humidity sensor. *Sensors and Actuators A: Physical* 298: 111593.

34 Naebe, M., Wang, J., Amini, A. et al. (2014). Mechanical property and structure of covalent functionalised graphene/epoxy nanocomposites. *Scientific Reports* 4: 4375.

35 Sahoo, S.K. and Wei, K.H. (2019). A perspective on recent advances in 2D stanene nanosheets. *Advanced Materials Interfaces* 6 (18): 1900752.

36 Wang, X., Pakdel, A., Zhang, J. et al. (2012). Large-surface-area BN nanosheets and their utilization in polymeric composites with improved thermal and dielectric properties. *Nanoscale* 7 (1): 662.

37 Rafiee, M.A., Narayanan, T.N., Hashim, D.P. et al. (2013). Hexagonal boron nitride and graphite oxide reinforced multifunctional porous cement composites. *Advanced Functional Materials* 23 (45): 5624–5630.

38 Zhang, X., Zhang, X., Yang, M. et al. (2016). Ordered multilayer film of (graphene oxide/polymer and boron nitride/polymer) nanocomposites: an ideal EMI shielding material with excellent electrical insulation and high thermal conductivity. *Composites Science and Technology* 136: 104–110.

39 Zhang, M., Wu, Q., Zhang, F. et al. (2019). 2D black phosphorus saturable absorbers for ultrafast photonics. *Advanced Optical Materials* 7 (1): 1800224.

40 Tran, V., Soklaski, R., Liang, Y.F., and Yang, L. (2014). Layer-controlled band gap and anisotropic excitons in few-layer black phosphorus. *Physical Reviews B* 89 (23): 235319.

41 Qiao, J.S., Kong, X.H., Hu, Z.X. et al. (2014). High-mobility transport anisotropy and linear dichroism in few-layer black phosphorus. *Nature Communications* 5: 4475.

42 Zhang, G.W., Huang, S.Y., Chaves, A. et al. (2017). Infrared fingerprints of few-layer black phosphorus. *Nature Communications* 8: 14071.

43 Wu, F., Xie, A., Sun, M.X. et al. (2017). Few-layer black phosphorus: a bright future in electromagnetic absorption. *Materials Letters* 193: 30–33.

44 Li, L.K., Yu, Y.J., Ye, G.J., and Ge, Q.Q. (2014). Black phosphorus field-effect transistors. *Nature Nanotechnology* 9 (5): 372–377.

45 Jain, A. and McGaughey, A.J.H. (2015). Strongly anisotropic in-plane thermal transport in single-layer black phosphorene. *Scientific Reports* 5: 8501.

46 Huang, L.N., Chen, C.G., Li, Z.J. et al. (2020). Challenges and future perspectives on microwave absorption based on two-dimensional materials and structures. *Nanotechnology* 31 (16): 162001.

47 Hanlon, D., Backes, C., Doherty, E. et al. (2015). Liquid exfoliation of solvent-stabilized few-layer black phosphorus for applications beyond electronics. *Nature Communications* 6: 8563.

48 Yuan, K.D., Wu, J.J., Liu, M.L., and Zhang, L.L. (2008). Fabrication and microstructure of p-type transparent conducting CuS thin film and its application in dye-sensitized solar cell. *Applied Physics Letters* 93 (13): 132106.

49 Miao, Q., He, D.W., Han, X.X. et al. (2020). Spatiotemporally resolved optical measurements on photocarrier dynamics in copper monosulfide. *Journal of Physical Chemistry C* 124 (27): 14459–14464.

50 Liu, P.B., Huang, Y., Yan, J. et al. (2016). Construction of CuS nanoflakes vertically aligned on magnetically decorated graphene and their enhanced microwave absorption properties. *ACS Applied Materials & Interfaces* 8 (8): 5536–5546.

51 Yaghi, O.M., O'Keeffe, M., Ockwig, N.W. et al. (2003). Reticular synthesis and the design of new materials. 423 (6941): 705–714.

52 Wang, L., Wen, B., Bai, X.Y. et al. (2019). Facile and green approach to the synthesis of zeolitic imidazolate framework nanosheet-derived 2D Co/C composites for a lightweight and highly efficient microwave absorber. *Journal of Colloid and Interface Science* 540: 30–38.

53 Leong, C.F., Usov, P.M., and D'Alessandro, D.M. (2016). Intrinsically conducting metal–organic frameworks. *MRS Bulletin* 41 (11): 858–864.

54 Espallargas, G.M. and Coronado, E. (2018). Magnetic functionalities in MOFs: from the framework to the pore. *Chemical Society Reviews* 47 (2): 533–557.

7

Conclusion and Perspectives

Two dimensional (2D) graphene has earned great attention from research community because of its outstanding electromagnetic interference (EMI) shielding ability. However, scalable production of high-quality graphene still remains a challenge. The use of common chemical or thermal routes creates many defects in the graphene lattice, significantly impairing the electrical properties. The electrical conductivity can be improved by carefully controlling the synthesis process or using less harsh experimental conditions. Incorporation with magnetic fillers and a controlled morphology can synergistically promote dielectric and magnetic polarization losses, which assist in electromagnetic wave (EMW) absorption. It is also important to understand how graphene is used for actual applications. Graphene films are flexible and mechanically strong and can be shaped into complex forms. For outdoor applications, graphene films can be sandwiched between polymeric films to avoid direct exposure to moisture and abrasion. Adhesive tapes should be developed upon which graphene films can be attached for customized shielding applications. Thin films are attractive for many applications, including portable electronics and telecommunication devices. Graphene films with nano- to micrometer thicknesses are a good alternative to metals traditionally used in these applications. So far, different methods have been used to develop thin films from graphene dispersions, including spin coating, vacuum filtration, solution casting, and polymer-assisted transfer methods. To meet large-scale demands, the roll-to-roll production of graphene film is important. Significant progress has already been made in this direction, but there is a need to control the properties from batch to batch.

More recently, transition metal carbides/nitrides/carbonitrides (MXenes) have emerged as a new class of 2D materials with superior properties as compared to graphene and many other materials produced to date. EMI shielding materials must possess good electrical conductivity, and MXenes fulfill this criterion by providing outstanding electrical conductivities. Electrical conductivity, as high as $15\,000\,\mathrm{S\,cm^{-1}}$, has been reported, which is a hundred times higher than that reported for chemically synthesized graphene. Additionally, owing to the excellent solution processability, MXenes have been developed in various forms including thin films, foams, polymer composites, and hybrid composites with other nanomaterials. MXenes also have the unique advantage of forming stable dispersions

Two-Dimensional Materials for Electromagnetic Shielding, First Edition.
Chong Min Koo, Pradeep Sambyal, Aamir Iqbal, Faisal Shahzad, and Junpyo Hong.

in common solvents such as water, making them suitable for spray coating. Thus, large areas can be shielded with MXene coatings in a short time. Tunable surface chemistry and the use of other MXene candidates with different elemental compositions or via solid solution approach can provide numerous opportunities for material design. Research on MXene-based shielding materials is moving at a fast pace, and various shortcomings, such as poor oxidation stability, and high synthesis cost, are expected to be addressed in the near future.

Various other 2D materials such as transition metal sulfides, nitrides, phosphides, a large family of transition metal dichalcogenides (TMDCs), and elemental 2D sheets are also discovered. Although these materials do not possess the outstanding electrical conductivity achieved by graphene or MXenes, they can create strong polarization losses that are beneficial for EMW absorption.

2D materials are at the forefront of materials research for EMI shielding because of their unique physical and chemical properties including high electrical conductivity, 2D sheet morphology, large surface area, abundant surface functional groups, corrosion resistance, and good solution processability. These advantages make 2D materials ideal candidates for shielding materials in different forms such as thin films, foams, polymer composites, and hybrid composites. 2D materials comprise a large family, including graphene, MXenes, TMDCs, black phosphorus (BP), hexagonal boron nitride (h-BN), and metal–organic frameworks (MOFs), thus providing much greater elemental choice than many other materials. Additionally, the construction of composites and hybrid structures of 2D materials with polymers or other functional nanomaterials provides ample opportunities for material design, which is set to advance EMI shielding materials research in the coming years.

Index

Two-Dimensional Materials for Electromagnetic Shielding, First Edition.
Chong Min Koo, Pradeep Sambyal, Aamir Iqbal, Faisal Shahzad, and Junpyo Hong.